INFORMATION
PROCESSING IN
BIOLOGICAL SYSTEMS

Studies in the Natural Sciences

A Series from the Center for Theoretical Studies
University of Miami, Coral Gables, Florida

Orbis Scientiae: Behram Kursunoglu, *Chairman*

A Continuation Order Plan is available for this series. A continuation order will bring
delivery of each new volume immediately upon publication. Volumes are billed only upon
actual shipment. For further information please contact the publisher.

INFORMATION PROCESSING IN BIOLOGICAL SYSTEMS

Chairman

Behram Kursunoglu

Editors

Stephan L. Mintz

Florida International University
Miami, Florida

and

Arnold Perlmutter

Center for Theoretical Studies
University of Miami
Coral Gables, Florida

PLENUM PRESS • NEW YORK AND LONDON

Library of Congress Cataloging in Publication Data

Orbis Scientiae (1983: Center for Theoretical Studies, University of Miami)
    Information processing in biological systems.

    (Studies in the natural sciences; v. 21)
    "Proceedings of the second half of the 20th annual Orbis Scientiae, dedicated to
P. A. M. Dirac's 80th year, held January 17–21, 1983, in Miami, Florida"—T.p. verso.
    Includes bibliographical references and index.
    1. Information theory in biology—Congresses. I. Kursunoglu, Behram, 1922–   . II.
Mintz, Stephan L. III. Perlmutter, Arnold, 1928–   . IV. Dirac, P. A. M. (Paul Adrien
Maurice), 1902–   . V. Title. VI. Series.
QH507.O73 1983   574.19   85-12160
ISBN 0-306-42075-9

Proceedings of the second half of the 20th Annual Orbis Scientiae
dedicated to P. A. M. Dirac's 80th year,
held January 17–21, 1983, in Miami, Florida

©1985 Plenum Press, New York
A Division of Plenum Publishing Corporation
233 Spring Street, New York, N.Y. 10013

Printed in the United States of America

The Editors Dedicate this Volume
to the Memory of
Paul Adrien Maurice Dirac
(1902–1984)

PREFACE

This volume contains the greater part of the papers submitted to the Information Processing in Biology portion of the 1983 Orbis Scientiae, then dedicated to the eightieth year of Professor P.A.M. Dirac. Before the volume could be published, Professor Dirac passed away on October 20, 1984, thereby changing the dedication of this volume, and its companion, on High Energy Physics, to his everlasting memory.

The last Orbis Scientiae (as it was often in the past) was shared by two frontier fields - in this case by High Energy Physics and Information Processing in Biology, demonstrating the universality of scientific principles and goals. The interaction amongst scientists of diverse interests can only enhance the fruitfulness of their efforts. The editors take pride in the modest contribution of Orbis Scientiae towards this goal.

It is a pleasure to acknowledge the typing of these proceedings by Regelio Rodriguez and Helga Billings, and the customary excellent supervision by the latter. The efficient preparation and organization of the conference was due largely to the skill and dedication of Linda Scott. As in the past, Orbis Scientiae 1983 received nominal support from the United States Department of Energy and the National Science Foundation.

The Editors
Coral Gables, Florida
April, 1985

CONTENTS

INFORMATION PROCESSING IN THE CORTEX: THE ROLE OF SMALL ASSEMBLIES OF NEURONS*

Gordon L. Shaw and John C. Pearson

University of California

Irvine, California   92717

ABSTRACT

Arguments are given for the proposed major roles played in cortical function by small assemblies of as few as 30 neurons. Predictions include that the processing capabilities of the primary visual cortex are much greater than those discovered in the pioneering work of Hubel and Wiesel. Results from experiments recording response from cat primary visual cortex to time sequences of bar stimuli show dramatic phenomena.

I.   INTRODUCTION AND ARGUMENTS FOR THE SMALL ASSEMBLY

A most important problem in understanding processing of sensory input in the mammalian cortex, as well as other tasks such as memory storage and recall, is the functional organization of groups of neurons. We have proposed[1] that there is a key mode of organization (similar to that of Mountcastle[2]) in the cortex characterized by small assemblies consisting of as few as 30 ("output" pyramidal) neurons (with perhaps a similar number of "local" interneurons). Major predictions are that cortical sensory processing involves flow of firing activity among assemblies as well as back and forth among

---
*Supported in part by the UCI Focused Research Program in Brain Function

different cortical areas, and that the "readout" or "coding" of

sensory cortical neurons involves the detailed structure[3,4] of their

spike trains and not simply the average firing response over some

substantial time step (~ 50-100 ms). In particular, we predict that

the processing capabilities of the primary visual cortex are much

greater than those discovered in the classic, pioneering work of

Hubel and Wiessel.[5] To examine these problems and test these predic-

tions, we developed a new experimental approach using detailed time

sequences of stimuli for the primary visual system (readily general-

ized to other visual areas of sensory systems[6]) which is powerful

even when recording from one microelectrode. Our recent, preliminary

results[7] from cat visual cortex are dramatic and demonstrate that

this approach will provide a "laboratory" for studying cortical net-

work behavior.

We begin by summarizing some of the arguments[1,2,4,8,9] for the

small assembly. In essence, we examine the next scale size in

cortical neuronal organization up from the single neuron. The con-

cept of (large) assemblies of neurons providing the basis for (a)

statistical reliability, (b) safety factor against local damage

(e.g., ~ 10^4 neurons "die" daily in human cortex and are not

replaced), (c) maintenance of firing activity for some fraction of

a second in the network, and (d) enhanced signal to noise has a long

history which includes important discussions by Lorente de No[10] and

Hebb[11]. We propose that these assemblies have a minimum (depending

on locus and function) of ~ 30 neurons which provide the basis for

cooperative phenomena not present in the individual neurons.

The classical access to neural circuitry has been through the

use of the Golgi staining procedures which allow visualization of 1

to 5% of the elements present in tissue scarcely 0.1 mm thick.

Figure 1a looks similar to a circuit wiring diagram. However, more

realistic representations of neural connections are becoming avail-

able through the use of scanning electron microscopy (SEM), providing

an incredible three-dimensional setting. In Figure 1b, the neurons,

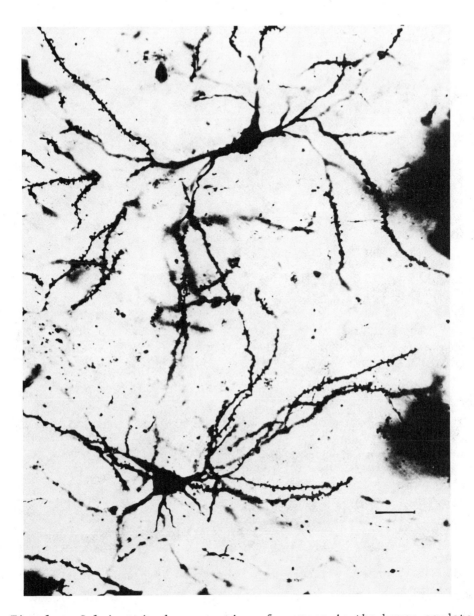

Fig. 1a. Golgi stained preparation of neurons in the human caudate nucleus showing cell bodies and segments of dendrites. Original magnification ×450.

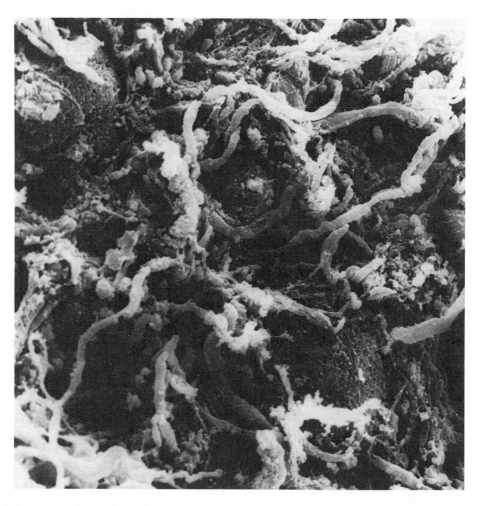

Fig. 1b. Scanning electron microphotograph of neurons in the rat
spinal cord immersed in neuropil. Original magnification ×2000.
These photographs are from the lab of A. B. Scheibel (Ref. 1).

dendrites, axons and glia appear strewn almost at random. (note
that the input synaptic connections, ~ 10^4 per neuron, on the neurons
and their dendrites cannot be seen at this scale). It is with this
new visual insight gained from the SEM that we review the role of
small assemblies of neurons in brain function. Although none of the
arguments is more than qualitative or suggestive, their diverse
origin (theoretical, anatomical and physiological) and combined
weight is substantial:

 i) Poststimulus histograms (PSH). Consider the presentation
of a "meaningful" stimulus to an animal that then has some "repeatable" behavioral response. It is well known that, although the
spike train response of a single cortical neuron is not reproducible[12], the PSH or (experiment) averaged response of a single
neuron to many presentations (roughly 10 to 40) of the stimulus to
the animal is "reproducible". If this widely used PSH is to have
physiological significance, we expect the firing response of a "network" containing the individual neuron to be repeatable because the
behavioral correlates to a single stimulus presentation are repeatable (as in the eye-blink paradigm of Thompson[12]). We assume that
the network is divided into assemblies of neurons defined, as in
Fig. 2, so that the firing response (to a single stimulus presentation) of the assembly averaged across the localized group of neurons
in it would be the same as the PSH of a single neuron (of a given
type) in the assembly. This definition for the assembly directly
relates the number of constituent neurons to the number of presentations necessary to achieve a "reliable" PSH, i.e., roughly 10 to 40.

 ii) Dendritic bundles. The presence of dendritic bundles (see
the SEM picture in Fig. 3) has been well established anatomically
in a number of neuronal systems throughout the mammalian nervous
system by the work of many investigators.[13] Dendritic bundles are
groups of dendrite shafts which course in very close apposition,
typically, from 1 μ apart to direct membrane apposition. With wide
variation, most bundles include 10-30 dendrites with bundle diameters

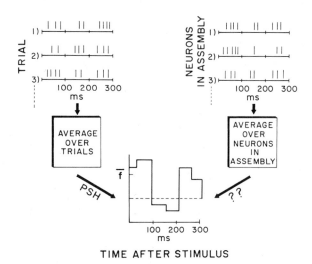

TIME AFTER STIMULUS

Fig. 2.   Schematic drawing illustrating our definition of an assembly
of neurons. The spatial average across neurons in the assembly for
one trial is assumed equal to (denoted by question marks) the average
for different experimental trials or poststimulus histogram (PSH)
for each individual neuron in the assembly. The upper left part of
the drawing represents the spike train responses of a typical
cortical neuron to repeated brief "meaningful" stimulus to the
animal. They vary from experimental (numbered) trial to trial.
However, the average $\bar{f}$ over trials or PSH (~ 10 to 40 trials) in the
lower center is "reproducible". The dashed line represents the background activity of this cell. The "size" of the time bins and the
relevant "levels" of average firing $\bar{f}$ will probably vary from region
to region in the cortex. The upper right part of the drawing
illustrates the spike train responses of the individual (numbered)
neurons in the assembly to one stimulus presentation.

of roughly 40 μ and a distance between bundles (in cortex) of ~ 60 μ.
Although few direct physiological experiments have been carried out
to determine their role, these bundles are unquestionably important
fundamental units in brain function. We infer this, for example,
from extensive work on bundles development showing[14] that, in many
regions, they are not present at birth. These anatomic units are
clearly of the appropriate size to be the basis for the assemblies
defined in i) and due to their close membrane apposition, could
carry out the direct electrical averaging needed for the theoretical
assemblies (also see iv)).

     iii) <u>Mountcastle's irreducible processing unit</u>. As detailed
by Mountcastle,[2] there are <u>many</u> data demonstrating that the cortex
is organized functionally into columns of roughly 500 μ diameter
(and 2000 μ in depth) consisting of ~ 10^3 - 10^4 vertically arranged
(perpendicular to the cortical surface), heavily interconnected
neurons (see, e.g. Fig. 1 of Lund[15]). The column is defined by its
inputs from the thalamus and its patterns of functional response.
For example, all the neurons in the column of the primary visual
cortex shown (in the <u>highly</u> idealized model) in Fig. 4 respond maximally to stimuli presented in the same locus in the visual field.
In his organizational principle for the functioning of the neocortex,
Mountcastle[2] regards the columns as the basic processing networks
(with interactions among columns). Each column consists of <u>irreduc-
ibly</u> small processing units: minicolumns of neurons, roughly 40 μ
in diameter. The minicolumns are then wired together into columns
or networks having the capability of performing its appropriate,
quite sophisticated, processing or memory tasks by being able to
exhibit sophisticated, dynamic firing patterns persisting for some
fraction of a second. The best studied, most highly organized
sensory regions are the <u>areas</u> of the visual cortex (see Fig. 5).
In particular, many properties of the primary visual cortex have been
determined in the pioneering work of Hubel and Wiesel.[5] The <u>highly</u>
idealized model of a column in Fig. 4 is divided into minicolumns

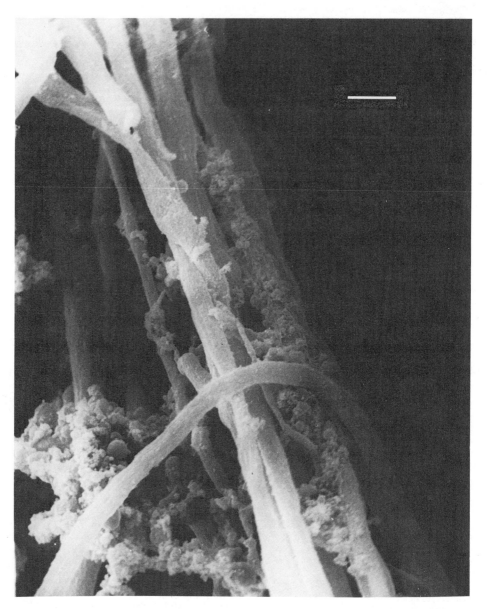

Fig. 3. Scanning electron microphotograph of a dendrite bundle in
the spinal cord of the rat. The bundle appears to contain five to
seven dendrite shafts, with elements entering or leaving along the
course of the bundle. Axons and neuroglia are also in contact with
the outer surfaces of the complex. Original magnification ×4000.
Scale bar 10 μ. From the lab of A.B. Scheibel (Ref. 1).

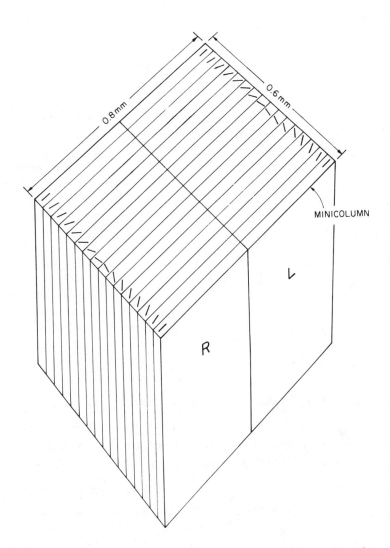

Fig. 4. Idealized model of a column in the primary visual cortex
with each minicolumn labeled by orientation (denoted by bar) and
ocular dominance. The firing response of the neurons in a particular
minicolumn is maximal to bar light stimuli presented in the orientations indicated and to the left (L) or (R) eye.

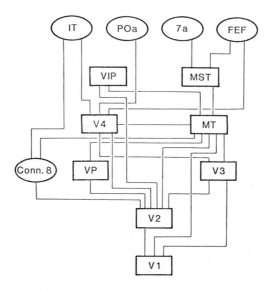

Fig. 5. Schematic outline of the hierarchical organization of the
many areas of macaque monkey cortex involved in visual processing.
Often overlooked is the fact that most of these connections are
reciprocal, with the higher areas feeding back to the more primary
regions. Thus, e.g. the "lower" area with its well known properties,
may serve as a window onto the activity of the higher areas. V1 is
the primary visual cortex (Fig. 4). This preliminary diagram is
from Van Essen and Maunsell (Ref. 27).

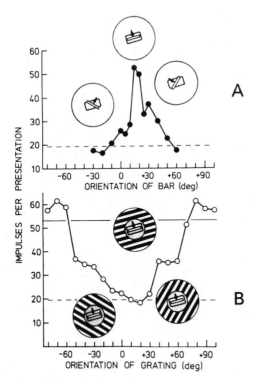

Fig. 6. (A) Standard orientation turning curve of a neuron in cat primary visual cortex. The larger rectangle represents the receptive field of the neuron (6.5° by 2.5°) and the smaller one is the moving bar. The ordinate shows the mean number of impulses (N=8) produced during a gating period of 2.5s centered on the time the stimulus was crossing the receptive field at a rate of 5 deg. s^{-1}. The dashed line is the mean spontaneous activity of the neuron. (B) Here we see the effect of "statistically" stimulating columns of cells which neighbor the column in which the monitored neuron lies. The response of the neuron to the moving bar at its optimal orientation is plotted as a function of the orientation grating (whose average luminance was was the same as the background in A). These data are from Blackmore and Tobin (Ref. 26).

in which the neurons will yield the maximum number of spikes when
the animal is stimulated by bar stimuli (stationary or moving
perpendicular to the orientation) which have a given orientation and
are presented to the left or right eye. (See the orientation tuning
curve in Fig. 6A). Note that the minicolumn dimension correspoding
to orientation is compatible with the Mountcastle 40 μ diameter and
our other estimates in this Section, whereas the long dimension is
an order of magnitude too large to identify with the "minimum"
size assembly. Thus, we suggested that an additional (functional)
"label" for the bar light stimuli should be present. Such an additional "label corresponding to "local" spatial frequency has been
reported.[16]

   iv) Processing or storage capacity of networks. A model
presented by Little and developed by Little and Shaw[17] mapped the
neural problem onto a generalization of the Ising spin model.
Briefly, the analogy goes as follows. Consider a network (or cortical column) of $N(\sim 10^3)$ highly interconnected neurons. We picture[17]
the firing (S = +1) or not firing (S = -1) at a given time and look
at the evolution of such patterns in discrete time steps, τ, of the
order of a few milliseconds, which is the order of the refractory
period and also the decay time of a postsynaptic potential. We
associate a spin 1/2 particle with each of the N neurons and assign
spin "up" ("down") when the neuron has fired (not fired) in the time
step. A configuration of all the spins then gives the firing pattern
of the network at a given time. A signal from a fired neuron propa-
gates along the axon to synaptic junctions on other neurons, thereby
influencing the probability of these neurons firing. As a result, a
new firing pattern will evolve which again can be described by a set
of N spins. Thus, in analogy with the Ising spin system in physical
problems, we can say that there is an effective spin-spin interaction
between sets of spins (neurons) in adjacent time steps. A function
analogous to the partition function of the spin system describes the
time evolution of the firing patterns of the neural network.

An important element of the model is that it is not deterministic. It is known that there are fluctuations in the postsynaptic
potentials due to the statistical nature of the release of the
chemical transmitter.[18] This indeterminacy gives rise to noise in
the network which is directly analogous to thermal noise in the spin
system and thus an effective "temperature" can be defined for the
network.[17] In spite of the noise, well defined states of the system
can occur which are analogous to clearly defined phases, like the
ferromagnetic or antiferromagnetic phases of material systems which
exist, of course, at finite temperature in the presence of thermal
noise. We emphasize that no true phase transitions are needed,
however, in the neural model (the ordered state in the physical system being analogous to the persistence of firing in the neural
network) because it is only necessary to maintain the firing for
~ 10 to 10^2 steps upon excitation.

It is generally assumed that the synapse is the most probable
place where change (plasticity) related to learning or memory or
processing capability takes place. The most widely used assumption
is the Hebb hypothesis[11] that the modification of synaptic strength
is dependent on correlated pre-post synaptic neuronal firing.
Although no experimental verification of this Hebb hypothesis has
been found, numerous theoretical studies have used it in studying
networks of neurons. Despite this enormous theoretical effort, the
large processing capability and large memory storage capacity of
mammals remain mysteries. We suggest that a new organization principle must be involved, and that Mountcastle's model[2] is a viable,
testable candidate for this principle.

Little and Shaw[17] using the Hebb hypothesis in their Ising
analogy model investigated the storage capacity of the network. For
a highly interconnected network of N neurons, there are $~N^2$ pieces
of information that can be encoded in the N^2 modifiable synaptic
strengths V. As noted, we picture the 2^N firing pattern α of the
network in terms of each neuron firing or not firing. We calculate

a $2^N \times 2^N$ probability matrix P, in terms of the N^2 V's, that a given firing pattern α goes to α' one time step later. In principle, one could determine P from the input-output firing patterns of the network and then decipher the synaptic strengths (or information) from it. Clearly, this cannot be easily done as there is enormous redundancy in the 2^N α's. (There are $2^N \times 2^N$ firing transitions in P, but only N^2 V's; thus, there is much redundant information in P.) Little and Shaw determined the independent, orthogonal memory "traces" by obtaining an exact solution to the linearized (large synaptic fluctuation) form of the model. They found that of the 2^N possible firing patterns α, only N specific linear combinations ϕ^a determine the network firing behavior. From these N ϕs, form N^2 transitions $\phi^a \rightarrow \phi^b$ which can be learned. Thus, a $2^N \times 2^N$ problem had been reduced to one of $N \times N$. However, the N mathematical ϕ's are linear combinations involving all 2^N α's, whereas the firing response of the network at a given time step is a specific α. Thus, there was no direct physiological interpretations of these solutions. Then, by introducing a Mountcastle-like organizational substructure, Roney and Shaw[19] obtained a self-consistent fully interpretable, predictive theory. The result is a network of 10^3 - 10^4 highly interconnected neurons which is divided into <u>assemblies</u>, (as defined in i) above) consisting of <u>roughly</u> 30 ("output") pyramidal neurons (with perhaps a similar number of local interneurons). Upon stimulation, activity flows from assembly to assembly, presumably persisting for some fraction of a second. A modified version of Hebb's learning hypothesis is obtained in which changes in individual synaptic strengths V are dependent on correlations of firing between the assembly in which the neurons providing the presynaptic input belongs and the assembly in which the postsynaptic neuron is a member. This might explain why previous experiments have failed to verify the original Hebb hypothesis.

v) <u>Reliability of assembly firing</u>. Reliability is defined here as the ability of the small assembly to sustain firing activity

(despite fluctuations) through at least 50 of our discrete time
steps after an initial excitation, yielding hysteresis curves very
similar to the magnetic analogy. The numerical results[20] depend
very little on the particular combinations of parameters (thresholds,
synaptic strengths, etc.), and we find that an assembly of as few
as 30 neurons will exhibit cooperative behavior.

   vi) <u>Sets of equivalent cells</u>. Bullock[21] who asks the
interesting question of how many "classes" or "sets" of essentially
"equivalent" cells (where the neurons in an equivalence class cannot
be distinquished) there are as a function of species. His estimated
distributions, given in Fig. 1 of Ref. 21 (or Fig. 3 of Ref. 1),
show a broad peaking in mammals at ~30 neurons per equivalence class.

   vii) <u>Functionally degenerate groups of cells</u>. Edelman[8] in his
selective theory of early brain development contends that outside
stimuli serve to select among preexisting configurations of cell
groups in order to create an appropriate response. Thus, <u>many</u>
cell groups can initially or in early development "recognize" a
given signal. Edelman estimates that his basic functional unit or
cell group could have as few as 50 neurons.

   viii) <u>Variability in neuronal activity correlated with reaction
response</u>. In a manner similar to that described in i), the reliability
or reproducibility of neuronal activity averaged over a group of
neurons in a particular paradigm is correlated with reaction time
to motor response in cat. The authors estimate[9] that their group
average involves 8-70 neurons.

   ix) <u>Synchrony of cortical-cortical inputs producing neuronal
firing</u>. Abeles stresses the importance of synchrony of inputs to
cortical cells from other cortical neurons in producing firing. One
cortical neuron <u>cannot</u> cause another cortical neuron to fire; on
the average, a cortical neuron has input synapses from $>10^3$ other
neurons. Abeles estimates that roughly 40 synchronous (to within
the average decay time, 2-10 ms, of postsynaptic potentials) inputs
to a pyramidal cell can fire it compared with roughly 300

"nonsynchronous" inputs. (Note that these synchronous inputs should
be considered as additional to an average background membrane
potential produced by nonsynchronous inputs.)

This summarizes the diverse arguments for the role of the small
assembly of neurons in brain function. (Note that previous
arguments[11,23] for assembly behavior estimated orders of magnitude
larger than 30 neurons). Points iii) and iv) lead us to the main
thrust of this paper, how to test for the dynamics of assembly
behavior. We predicted[24] that the processing capabilities of our
basic network or cortical column (involved dynamic interactions
among assemblies or minicolumns, see Fig. 4) and are much greater
than presently documented. In particular, processing of rotational
stimuli in the primary visual cortex was suggested. Motivated by
this, a psychophysics experiment was conducted[25] that demonstrated
a spatial-temporal filling-in-process in apparent motion. We then
presented this "human illusion" to cats and recorded from neurons
in the primary visual cortex with results that demonstrate striking
network phenomena: Stimuli were presented to cats maintained under
N_2O and flaxidil. Microelectrodes were placed in area 17 with the
aid of stereo-taxic coordinates and the spike firing response of
"groups" of a few (2-4) neurons (separated by spike amplitude from
background) were observed.[7] The center of the visual field and
optimum bar orientation were determined for the monitored group
(see Fig. 6A), and then repeated time sequences corresponding to
"clockwise" and to "counterclockwise rotations" of the stimulus
orientation. Although the "filling in" process has not yet been
clearly observed, several striking features have been found. For
example: a) We observed dramatic sharpening as well as greatly
increased peak firing response when the optimum stimulus is left
out of the sequence (for one group, the width of the poststimulus
histogram at the time of the second bar presentation decreased from
30 ms to leass than 4 ms). b) We noted a striking change in firing
activity dependent solely on the change of the sense of presentation

from "clockwise" to "counterclockwise". The details of these experiments will be summarized in the next Section.

These results lead us to the crucial questions concerning the nature of coding of information in neuronal processing. As noted by Segundo,[3] there are at least two important, necessary, but not sufficient requirements for particular observed neural firing patterns to be part of the animal's "natural" coding: These observed neuronal firing patterns should occur in normal behavior; and they must be able to effect firing of other (efferent) neurons (i.e., not lead to a "dead end"). We believe that our simple, time-dependent stimuli, related to apparent motion would "occur" in nature, and that some of our observed neuronal firing responses could have large effects on their efferent neurons. In particular, the dramatic decrease in width of the peaks noted above of the post stimulus histogram response to certain stimuli (see Fig. 7A) could greatly effect other neurons: If, indeed, these PSH's represent an enormous increase of synchronous output of such neurons. By Abeles' arguments in point ix), they become extremely efficient in their ability to fire their efferent neurons.

Some further speculations are made in the final section.

II. NEURONAL FIRING RESPONSE IN PRIMARY VISUAL CORTEX TO SEQUENTIAL
     PRESENTATION OF LIGHT BAR STIMULI

The dynamics in cortical sensory processing involves firing activity among large numbers of neurons. We wish to study this enormously important problem by building on the work of Hubel and Wiesel[5] who found the remarkable "feature detector" properties of responses of single neurons in the primary visual cortex. The great advance by Hubel and Wiesel in understanding visual information processing depended on their use of visual stimuli that are elements of real-world objects such as lines or edges in contrast to flashes of light used by previous workers. However, these bars shown in isolation will not reveal the full network properties and capabilities of the visual cortex. Thus, our approach has been to present

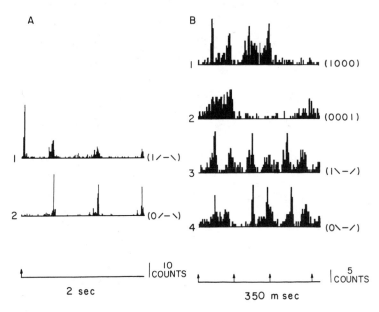

Fig. 7. (A) Poststimulus histograms of group 1 in two different stimulus sequences. The first angle presented in each cycle had been previously determined to be the optimum orientation. The following three angles were inclined at 45°, 90°, or 135° from this angle. One cycle of the stimulus is described by a sequence of lines whose orientations are relative to the optimum angle. A vertical line always denotes the optimum angle. Thus, a clockwise rotation is described by |/ — \ , and a counter-clockwise rotation is described by |\ — / . A 0 is such a sequence denotes that no bar presented at that point in the cycle, and a ? denotes that one of the nonoptimum angles was randomly selected to be presented at that point. Each stimulus was on for 500 ms and off for 150 ms. Bin width has 4 ms and the number of cycles was 30. The arrow indicates the time of onset of the first stimulus. (B) Poststimulus histograms of group 2 in four different stimulus sequences. Each stimulus was on for 32 ms and off for 70 ms. The bin width was 3 ms; and the number of cycles was 30. The arrows indicate the time of onset of each stimulus.

these stimuli in time sequences with the aim of <u>sequentially</u> exciting neurons consistent with the organization of the primary visual cortex determined by Hubel and Wiesel and involving the neuron being monitored. Our "third generation" preliminary studies[7] have already yielded striking, predicted and unpredicted phenomena.

As summarized in Fig. 4 for the <u>highly</u> idealized model of a cortical column in the primary visual cortex, we know (from Hubel and Wiesel, etc.) the nature of localized bar stimuli that maximally excite neurons in each minicolumn. Suppressing other labels identifying a bar and a minicolumn by their orientation, e.g., the bar | will excite the micicolumn |. The response of the column to the stimulus | is the "static" firing pattern in the sense that the activity is confined to the one minicolumn |. However, a repeated time sequence of bar stimuli presented in discrete successive orientations centered about the same point in space corresponding to a clockwise |/ —— \ (or a counterclockwise |\ —— /) sense should excite a <u>flow</u> of activity in the minicolumns |/ —— \ since these minicolumns have adjacent cortical locations. We expect distinctive network firing behavior, <u>not</u> simply determined from the "static" responses to the individual stimuli. For example, we predicted[24] from our theoretical studies[19] (see point iv) in Section 1) that omission of one bar in these discrete time sequences of "rotating" stimuli (e.g.,|/ 0 \ where 0 denotes the absence of a stimulus) when presented at the appropriate speed, would be filled in at the neuronal level. Also, we anticipated <u>different</u> single neuron response to a clockwise versus counterclockwise sequence. We expect that such time dependent sequences constitute the major processing capabilities of the column and that the "static" ones observed thus far constitute a subset of the possible patterns capable of being excited.

We designed and performed a comparable human psychophysics experiment[25] in which an observer was presented (via graphics on a microcomputer) the repeated time sequences a) |/ —— \, b) |/ 0 \

c) |/ - \ where the 0 in b) denotes the absence of a stimulus and - in c) denotes a segment of bar ——. For a presentation rate of roughly .7 hz, all three sequences appeared as continuous rotations of the bar. Using polarizing material to have half of the stimulus enter one eye and half in the other, we have shown that this filling in illusion must take place at the cortical level since binocular excitation of neurons first occurs there.

A crucial question addressed here is whether rotation is indeed processed in the primary visual cortex and whether filling in "illusion" could be observed there at the single neuron level. Stimuli were presented to cats maintained under N_2O and flaxidel. Extracellular microelectrodes were placed in the primary visual cortex with the aid of sterotaxic coordinates and groups of a few neurons were observed. The center of the visual field and optimum bar orientation were determined for the monitored group, and then repeated time sequences corresponding to monitored group, and then repeated time sequences corresponding to clockwise and to counter-clockwise rotations of the stimuli were presented, both _with_ and _without_ the optimum stimulus orientation. Although the "filling in" process has not as yet been clearly observed, several striking features have been found.[7] Five of these phenomena are shown in Fig. 7 and 8:

i. Sharpening and increased peak response: A dramatic sharpening as well as _greatly_ increased peak firing response are noticed in Fig. 7A when the optimum stimulus is left out of the sequence. The width of the post-stimulus histogram at the time of the second bar presentation decreases from ~ 30 ms in Fig. 7A.1 to less than 4 ms in Fig. 7A.2. Note that this sharpening occurs but not as much for the other angles. In addition, the background firing is suppressed.

ii. Dependence on order of presentation: Note the striking change from Fig. 8.2 to Fig. 8.4 dependent solely on the change of the sense of presentation |/ —— \ to |\ —— /, i.e., from

"clockwise" to "counterclockwise". (This is also clearly evident
in our recent unpublished <u>single</u> neuron data.[7])

   iii. <u>Suppression of secondary peaks</u>: Some groups showed two
activity peaks (Fig. 8.1) or even four peaks (Fig. 7B.1) of large
size when only one bar orientation was shown. When a time sequence
of other orientations was shown, then some of the peaks were suppressed. This is clearly seen in comparing Fig. 8.1 and 8.2. Note
that the onset of the second bar presentation in Fig. 8.2 comes
<u>after</u> the time when the suppressed second peak response to the first
bar (Fig. 8.1) would be present. Thus, this suppression <u>cannot</u> be a
simple inhibition due to the onset of the next stimulus as might be
inferred in the case of Fig. 7B.

   iv. <u>Possible filling-in</u>. Figure 7B.4 might appear as evidence
for a filling in or firing of the group in <u>at the appropriate</u> time
even when the first bar is omitted. Caution must be used here,
however, due to the possible presence of secondary peaks from the
fourth bar.

   v. <u>Support for the assembly hypothesis</u>: In Fig. 8.6, we
present some neuronal spike response data which was <u>barely</u> above
the background noise and had been separated from the response shown
in Fig. 8.5 by spike amplitude discrimination. Note that the sharp
response to the sequence (o\ —— /) was remarkably similar for
these additional neurons, giving further support for our assembly
hypothesis (see Fig. 2).

   Finally we note from our more recent, preliminary, <u>single</u> neuron
data that varying the on and off times of the sequential stimuli
lead to large enhancement of some of the effects noted above. This
may indicate "resonant" flow of activity.

III. CONCLUSIONS AND FURTHER SPECULATIONS

   Although the striking results presented in Section 2 clearly
justify the usefulness and richness of our approach, <u>much more</u>
experimental and theoretical work needs to be done to further
establish, elaborate and understand these results. We believe that

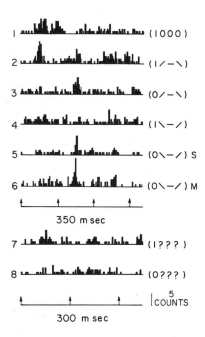

Fig. 8. Poststimulus histogram of group 3 in eight different stimulus sequence. Each stimulus was on for 32 ms and off for 70 ms
in sequences 1 through 6 and 90 ms in sequences 7 and 8. The bin
width was 3 ms and the number of cycles was 20. In sequences 7 and
8 only the onsets of the first three stimuli are marked. Sequences
5 and 6 were derived from the same recording but include the discharges of different neurons. Sequence 5 is composed from the same
small group (S) of cells as the others in this figure, while
sequence 6 includes a larger group of neurons (M) whose discharges
were just above the noise level. These data are from Ref. 7.

our paradigm will open up a whole new area of research and many
other interesting new features will be found. For example, the
sequential stimuli discussed in Section 2 simulated rotation and
activated sequential orientation minicolumns in the same cortical
column. Our purpose is to elicit and understand the dynamical
firing interactions among minicolumns in the same "network." It
is well known, see Fig. 6B, that there are "static" interactions
among columns[26] in the primary visual cortex. We intend to examine
the dynamic time dependent interactions among columns in cat visual
cortex by sequentially presenting bars of light at the same orientation but at distances corresponding to the receptive field of
neighboring columns. The analogous psychophysics experiments[25]
in apparent linear motion have been presented to human observers.

       As noted earlier, there are many areas of the cortex involved
in processing visual sensory input in a complicated manner. Fig.
5 from the work of Van Essen and Maunsell[27] shows the "hierarchical"
organization of cortical visual areas in the macaque monkey. A
profound, often ignored fact is that there are as many connections
back from the higher (less retinotopic) areas as there are forward
connections! At first thought, this might seem to preclude deci-
firing the sensory "coding" involved in visual processing. However,
these feedback loops might enable us to understand some of the
"higher" areas by appropriate monitoring of the primary visual
cortex. That is, the primary visual cortex could provide a "window"
for examining the higher visual areas. For example, the response
sharpening in the primary visual cortex upon leaving the optimum
stimulus out of a time sequence, described in Fig. 7A, might occur
due to feedback.

       Carrying this discussion on feedback even further, we speculate
on the possible role the primary sensory areas play in thought or
processing information even in the absence of outside sensory input.
Perhaps, as suggested by Harth,[28] the primary areas act as "sketch
pads" so that, e.g., when a person closes his eyes and imagines a

rhino, a representation of a rhino "appears" in neuronal coding in the primary visual cortex. A powerful new technique, positron emission tomography[29] (PET) scanning now exists which could be used to test this intriguig idea in the auditory sensory system. PET scans are used in human subjects to monitor activity in the brain. The spatial resolution (~ .5 mm) is sufficient to do the following experiment: Consider a trained, professional musician who has a particular concerto in his repertoire. Having questioned several musicians, they all claimed that when rehearsing a concerto in their heads, they "heard" the music. Thus, we suggest PET scans of a musician's cortex when he is "listening" to a familar piece of music in his head with no outside input as compared to when he is listening to a recording to the same music. We suggest that the primary auditory cortex will appear similarly illuminated in both PET scans.

Next, we speculate on two possible roles of statistical fluctuations in brain function. We have concerned ourselves in Section 1 with how the brain is able to generate "reliable behavior" in the presence of considerable noise or fluctuations in postsynaptic potentials and spike train outputs, synaptic depression, and seemingly indeterminate local connectivity, as well as many other varying inputs. We have argued that the cooperativity phenomena exhibited by the small assemblies enable this reliable behavior to occur. However, we further believe[1] by physical analogy to the Ising spin system that both the ease and speed in switching from one neuronal firing pattern to another, and the large processing (or memory storage) capabilities of neural networks require fluctuations. In particular, to understand this conjecture concerning the processing capabilities we exploit the recent work by Fisher and coworkers[30] who consider an Ising model for a magnetic material which has three types of couplings between the spins depending on direction and distance of separation (ANNNI model). Now, in contrast to the simple Ising model with only one type of interaction which gives only one (simple) type of phase transition, they found infinitely many,

stable phases. Two important restrictions were found in their
analysis: (a) One of the three types of interactions must be repul-
sive and the other two attractive, (b) Many states exist only at
finite temperature. These striking results are shown in Fig. 9.
These two restrictions are significant for our neural network
analogy: restriction (a) translates to the requirement that there
must be inhibitory as well as (two types of) excitatory synapic
potentials V, and (b) indicates that fluctuations must be present.
We[31] have made considerable progress inpursuing this neural analogy
to the ANNNI model results.

We conclude by referring to the recent study[32] comparing rotational invariance in visual pattern recognition by pigeons and
humans. The fascinating results in Fig. 10 show that, in contrast
to the linear rise in human reaction time with rotation of the
comparison object, the pigeons' reaction time is independent of
rotation. Furthermore, the pigeons' reaction time is considerably
faster. This clearly demonstrates how nervous systems can evolve
to solve specific problems; this rotational pattern recognition
problem is more important to a flying animal than to one who walks.
Finally, it gives support to the belief that the understanding of the
principles involved in brain function will lead to as major a
revolution in the technology of information processing as the advent
of the electronic digital computer.

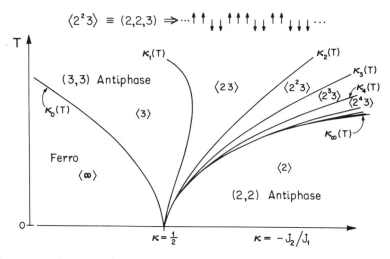

Fig. 9. Schematic phase diagram of the anisotropic next-nearest-neighbor Ising or ANNNI model in the plane of temperature T and parameter $K = -\dfrac{J_2}{J_1}$, exhibiting the infinite sequence of commensurate, layered anti-phase states, $\langle 2^{J-1} 3 \rangle$, at low temperatures. $\langle 2^{J-1} 3 \rangle$ means a sequence of J-1 pairs of lattice layers pointing (predominantly) two "up" and two "down," followed by three layers all pointing (predominantly) "up" or "down" to maintain an overall "anti phase" character as illustrated for $\langle 2^2 3 \rangle$ in the figure above. J_2 is the next nearest neighbor coupling between layers along the axis of anisotropy and is antiferromagnetic ($J_2 < 0$). J_1 is the nearest neighbor coupling between layers along the axis of anisotropy and is ferromagnetic ($J_1 > 0$). There is an additional ferromagnetic interaction ($J_0 > 0$) between nearest neighbor spins within a layer. These results are from Fisher and Selke, Ref. 30.

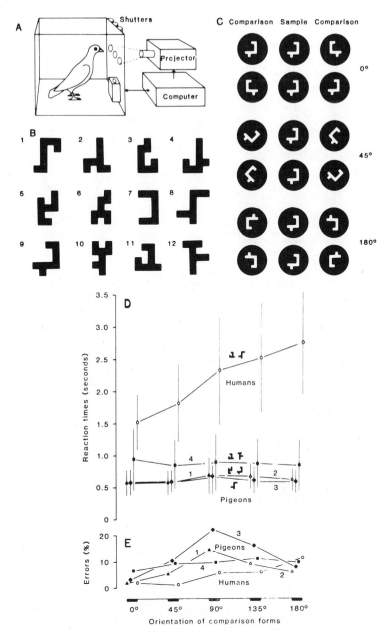

Fig. 10. (A) Experimental apparatus used to test pigeons' ability to choose which one of two alternative visual forms was identical to, or a mirror image of, a previously presented sample form. (B) Visual forms used in test. (C) Examples of stimulus sets used for rotational invariance test. Mean reaction times with standard deviations (D) and mean error rates (E) as a function of comparison form rotations. Data are from 9 pigeons and 22 humans. These data are from Holland and Delins, Ref. 22.

REFERENCES

1. G.L. Shaw, E. Harth, and A.B. Scheibel, Exp. Neurol. <u>77</u>, 324
    (1982).

2. V.B. Mountcastle, In G.M. Edelman and V.B. Mountcastle (Eds.)
    <u>The Mindful Brain</u>, MIT Press. Cambridge, MA (1979).

3. J.P. Segundo, p. 569, <u>The Neurosciences: Second Study Program</u>,
    Ed. F.O. Schmidt, MIT Press (1970).

4. M. Abeles, <u>Local Cortical Circuits</u>, Springer-Verlag, Berlin
    (1982); Israel J. of Med. Sci. <u>18</u>, 83 (1982).

5. D.H. Hubel and T.N. Wiesel, Proc. Roy. Soc. <u>B198</u>, 1 (1977); J.
    Physiol. (London), <u>160</u>, (1962); J. Physiol. (London) <u>165</u>,
    559 (1963).

6. Preliminary studies in the auditory cortex of cats are now being
    started in the laboratory of N.M. Weinberger (private communi-
    cation) using sequential step tones.

7. G.L. Shaw, P.C. Rinaldi and J.C. Pearson, Exp. Neurol. <u>79</u>, 293
    (1983); J.C. Pearson, D. Diamond, T. McKenna, P.C. Rinaldi,
    G.L. Shaw, N.M. Weinberger, Abstract for 13th Annual Meeting,
    Society for Neuroscience (1983).

8. G.M. Edelman, In, <u>The Mindful Brain</u>, (1979); p. 535, <u>The</u>
    <u>Organization of the Cerebral Cortex</u>, Eds. F.O. Schmitt, et. al.,
    MIT Press (1981).

9. J.F. Dormont, A. Schmied and H. Condé, Exp. Brain Res. <u>48</u>, 315
    (1982).

10. R. Lorente de Nó, J. Neurophys. <u>1</u>, 207 (1938).

11. D.O. Hebb, <u>The Organization of Behavior</u>, Wiley, N.Y. (1949).

12. B.D. Burns, <u>The Uncertain Nervous System</u>, Arnold Pub., London
    (1968); R.C. Dill, C.E. Vallecalle and M. Verzeano, Phys.
    Behav. <u>3</u>, 797 (1968); A. Ramos, E.L. Schwartz and E.R. John,
    Science <u>192</u>, 393 (1976); R.F. Thompson, Am. Psychol. 31, 209
    (1976).

13. See references in the review, K.J. Roney, A.B. Scheibel and
    and G.L. Shaw, Brain Res. Rev. <u>1</u>, 225 (1979).

14. M. E. Scheibel and A. B. Scheibel, p. 347 in Golgi Centennial Symp. Proc., Ed. M. Santini, Raven Press, New York (1975).

15. J. Lund, p. 105, The Organization of the Cerebral Cortex, Eds. F. O. Schmitt et al., MIT Press (1981).

16. R. B. Tootell, M. S. Silverman and R. L. de Valois, Science 215, 813 (1981).

17. W. A. Little, Math. Biosci. 19, 101 (1974); G. L. Shaw and R. Vasudevan, Math. Biosci. 21, 207 (1974); W. A. Little and G. L. Shaw, Behav. Biol. 14, 115 (1975); Math. Biosci. 39, 281 (1978).

18. B. Katz, Nerve, Muscle and Synapse, McGraw-Hill, New York (1966).

19. G. L. Shaw and K. J. Roney, Phys. Lett. 74A, 146 (1979); Math. Biosci. 51, 25 (1980).

20. E. Harth, N. S. Lewis and T. J. Csermely J. Theor. Biol. 55, 201 (1975); N. S. Lewis, Doctoral Thesis, Syracuse U. (1974); J. C. Pearson and G. L. Shaw, unpublished.

21. T. H. Bullock, p. 1 in Neurobiology of the Mauthner Cell, Eds. D. Faber and H. Korn, Raven Press (1978).

22. Clearly, this minimum size depends on the location in the brain of the network and its functions. For example, consider the olfactory bulb in which, (W. J. Freeman, Mass Action in the Nervous System, Academic Press, N. Y. 1975), only a few, extremely precise discriminations in smell are carried out. Here, presumably (and in agreement with the multiple micro-electrode recordings of Freeman) the assemblies should be large.

23. E. R. John, Science 177, 850 (1972).

24. G. L. Shaw, P. C. Rinaldi and J. C. Pearson, to be published.

25. G. L. Shaw and V. S. Ramachandran, Perception, in press.

26. C. Blackmore and E. A. Tobin, Exp. Brain Res. 15, 439 (1972).

27. D. Van Essen and J. Maunsell, private communications.

28. E. Harth, private communication.

29. See the review by M. E. Phelps, J. C. Mazziotta and

S.-C. Huang, J. of Cerebral Blood Flow and Metab. $\underline{2}$, 113 (1982).

30. M. E. Fisher and W. Selke, Phys. Rev. Lett. $\underline{44}$, 1502 (1980).

31. J. Patera, J. C. Pearson, G. L. Shaw and R. Vasudevan, unpublished.

32. V. Holland and J. Delins, Science $\underline{219}$, 804 (1982).

INFORMATION AND CAUSE

Robert Rosen

Dalhousie University

Halifax, Nova Scotia, B3H 4H7 Canada

I. INTRODUCTION

The study of biological systems revolves around but a single
basic task: to characterize how systems which are living are distinguished from all other systems, and to use this characterization to
determine how it is that organisms can do all the unique things they
do. In Schrodinger's words, then, the basic question of biology is
simply: what is life?

Theoretical physics has set for itself an apparently larger
task: that of the elucidation of physical nature in general. In the
pursuit of this task over the course of centuries, theoretical
physics has evolved into a marvellously subtle and sophisticated
endeavor, which indeed constitutes one of the supreme achievements
of man as a species.

Biology has played a negligible role in this evolution, either
as cause or as effect. If Wigner could speak of the unreasonable
effectiveness of mathematics as the language of theoretical physics,
one can also speak of the unreasonable inneffectiveness of the
relationship between theoretical physics and biology. It is appropriate on this occasion, which brings physicists and biologists
together in honor of one of the towering architects of modern physics,

31

to address this unreasonable situation, and see what may lie behind
it.

I think it fair to say that the professional theoretical
physicist ignores biology because it seems to him so irreducibly
contingent. That is, to a physicist organisms appear to be inordi-
nately special systems, as evidenced by the fact that only an
insignificant minority of material systems can be regarded as living.
Therefore, it seems inappropriate for the theoretical physicist, in
his quest for universal realities, to seek them in the organic realm;
it is more natural for him to regard organisms as remote consequences
of such universals, supplemented perhaps by many special constraints
and initial conditions. He does not regard it as his task to specify
these. So it is that, with a few exceptions made more notable by
their rarity, the world of organisms has played no part in the growth
of physics, either in theory or in practice.

Oddly enough, the vast majority of modern biologists avidly
embrace this perspective. In fact, they go much further. It having
been ascertained that the material constituents of organisms do not
violate physical universals, biologists accept as an article of faith
that "physical laws" are ultimately responsible for organic behavior.
But they also believe that organisms are so special that such laws
are irrelevant; indeed, a growing consensus is that the character of
organic life is primarily shaped by historical accident and, therefore, is immune to understanding in terms of logical necessity or
physical law. Thus, biology stands doubly estranged; in the
formulation of natural laws by the physicist, and in the application
of natural laws by the biologist.

Another view of this situation can be entertained, however; one
which has hardly been articulated, let alone explored. It is the one
which I would like to argue here: namely, that the apparent ortho-
gonality between theoretical physics and biology arises not because
physics is too general, but rather that it is too special. Specifically, I will argue that the almost total preoccupation of physicists

with non-organic systems (and in particular, with systems which are closed, isolated and conservative) has led them to a language in which certain basic features of physical reality have been systematically excluded. It is precisely these excluded features which are required if we are to discuss organic matter, and indeed matter in general, in full generality.

In exploring such a possibility, which has been so consistently ignored by physicists and biologists alike, we may recall the words of Albert Einstein: "One can best appreciate from a study of living organisms how primitive physics still is."

II. ON "INFORMATION"

Before embarking on our main theme, we will digress to provide a few words of motivation and historical perspective.

The estrangement between theoretical physics and biology, to which we have alluded above, is seen perhaps most sharply by comparing the vocabularies employed by the two disciplines. That of physics is dominated by words like state, force, energy, potential. Biology, when it uses these words at all, does so in a superficial and nontechnical way; its vocabulary is replete rather with words pertaining to organization and to information. Words like "program", "control", "computation", etc., simply have no physical counterpart, and are hardly ever used seriously by physicists. This divergence of vocabularies, more than anything else, underlies the feeling that there is simply no continuum over which the properties of organic and nonorganic systems can be related and compared.

However, the otherwise unrelated lexicons of physics and biology do have one point of contact, and that is in the circle of ideas revolving around the concept of stability. I have felt for a long time that this point of contact might provide the Rosetta Stone which would allow us to translate physical concepts into informational ones, and in this way to build an initial bridge between fundamental physics and biology.

Actually, my initial ventures in this direction have been far more

modest. It is well known that "information" is one of the murkiest
and most overworked words in the scientific lexicon (perhaps only
the word "model" can be compared to it). Indeed, there is almost
no relation between the "information" of the Information Theory
of Shannon (which beguiles many through its superficial relation to
entropy), the "genetic information" upon which development, physiology and evolution devolve in their several ways, and the
"information" appearing diversely in the functioning of the brain.
Thus, to relate physics to "information" from the outset seems far
too broad a task. Better perhaps to take certain relatively well-
defined concepts closely associated with the "processing" of
"information", and explore how these may be translated first into
dynamical language, and from there into the language of energies,
potentials and forces.

The terms chosen for this purpose were "activation" and its
inverse, "inhibition". These words appear over and over, in various
guises, in the study of genetic processes, of enzymatically catalyzed reaction networks, in developmental biology, in physiology,
and in the theory of the brain, always associated with "information
processing" and cybernetic control. Indeed, any biological activity
can (rather glibly) be regarded as the consequence of an interaction
between antecedent processes which facilitate it (i.e. activators)
and those which repress it (inhibitors).

The point of departure was a suggestion initially made explicit
by Higgins (1967); though as will be seen its antecedents go back
much earlier. What Higgins observed was the following: given a
system of first-order rate equations

$$dx_i/dt = f_i(x_1, \ldots, x_n), \quad i = 1, \ldots, n \qquad (1)$$

i.e., a dynamical system, of the type which forms the ubiquitous
starting point for discussing temporal behavior in all kinds of
real systems, the concept of activation/inhibition may be directly
characterized.

In particular, we think of x_1, ..., x_n as <u>state</u> <u>variables</u>
for our system; their values at instants of time appear as the
arguments in the dynamical equations (1), and define system <u>states</u>.
At any state, under relatively mild conditions, we can define the
n^2 quantities

$$u_{ij}(x_1, \ldots, x_n) = \frac{\partial}{\partial x_j}\left(\frac{dx_i}{dt}\right) \tag{2}$$

If such a quantity is <u>positive</u> at a state, it means that an increase
in the value of x_j in that state will increase the rate at which
x_i is changing (or alternatively, that a decrease in x_j will dimin-
ish the rate at which x_i is changing). It is, thus, natural to say
that x_j is an <u>activator</u> of x_i in the given state. Conversely, if
u_{ij} is <u>negative</u>, it is likewise natural to say that x_j is an <u>inhibi-</u>
<u>tor</u> of x_i in that state.

We showed some years ago that these quantities u_{ij} determine
what we called an <u>activation-inhibition network</u>, strongly remini-
scent of the kinds of "information-processing" networks which have
long been associated with the theory of the brain (cf. Rosen 1979).
In particular, let us associate a formal "element" with each state
variable in (1). This element is imagined to be equipped with n
afferent "lines" along which differential changes dx_j in the values
of the other state variables determine their "state of excitation".
Each such differential change is weighted by the appropriate value
of the factor $u_{ij}(x_1, \ldots, x_n)$. Thus, each of these formal
"elements" sees a total excitation given by the differential form

$$\sum_{j=1}^{n} u_{ij}dx_j = df_i \tag{3}$$

From (1), we see that this differential form is nothing but the
differential increment in the velocity dx_i/dt. The function of this
formal element is then to convert this velocity increment into an
increment of the associated state variable; i.e., to perform an

integration which we may symbolize as

$$df_i \longrightarrow dx_i.$$

The process is then repeated with the updated values of the state variables. The analogy with e.g. neural networks should be obvious.

Before proceeding further, it may be well to pause and contemplate a few consequences of this picture, which are of considerable interest in themselves, and which generalize to provide still more interesting consequences later.

First, we have seen that a system of rate equations (1) automatically gives rise to an activation-inhibition pattern (2). We may ask conversely whether the specification of an activation-inhibition pattern (2) can be converted into a system of rate equations (1). That is: are the dynamical language and the informational language equivalent?

The answer is obviously negative; the informational language is more general. In order to pass from (2) to (1), it is obviously necessary that all the differential forms (3) should be exact, which is clearly a highly non-generic situation. It follows that the perturbation of the activation-inhibition pattern of a dynamical system (1) is, in general, a far more drastic thing than is the perturbation of dynamics itself.

The conditions for the exactness of (3) are themselves instructive from the "information" point of view. The conditions are the equality of mixed partials:

$$\frac{\partial}{\partial x_k}\left(\frac{\partial}{\partial x_j}\left(\frac{dx_i}{dt}\right)\right) = \frac{\partial}{\partial x_j}\left(\frac{\partial}{\partial x_k}\left(\frac{dx_i}{dt}\right)\right) \tag{4}$$

These mixed partials can be interpreted as follows: the sign of the quantity

$$\frac{\partial}{\partial x_k} u_{ij}$$

measures the effect of a variation in x_k on the degree of activation
or inhibition of x_k by x_j; if this sign is positive, we may call
x_k an <u>agonist</u> of x_j, if negative, an <u>antagonist</u>. The conditions (4)
assert that the agonist/antagonist relation between x_j and x_k be
symmetric for each i; again, a highly non-generic situation.

Finally, if the forms (3) are inexact, it turns out that we do
not lose thereby all possibility of expressing an activation-
inhibition pattern in dynamical terms. What we lose is only the
ability to express the rates of change of the state variables (i.e.,
the velocities f_i) as functions of states. In effect, we must regard
our former state space as if it were the configuration space of a
mechanical system, and embed it in an appropriate phase space. We
can then write down the following equations of motion:

$$dx_i/dt = f_i \qquad\qquad i = 1, \ldots, n$$

$$df_i/dt = \sum_{j=1}^{n} u_{ij} f_j.$$

This necessity to pass to a <u>second-order</u> system is a first glimmering
of a relation between "information" (as embodied in an activation-
inhibition network) and force. This relation will become more
suggestive as we proceed.

III. ON "CAUSE"

The ideas sketched in the previous section are still very
special. In order to understand properly why they are so special
and how they may be appropriately widened, we must make a brief
detour into the epistemology of relations like (1) above. Epistemological considerations are as irksome to physicists (and biologists)
as foundational questions are to mathematicians, and for a similar

reason: let well enough alone. But we will find that they are essential, and that they quickly lead us into some remarkable relations between "information" and causality.

To fix ideas, let us look again at a relation like

$$v_i = dx_i/dt = f_i(x_1, \ldots, x_n) \ .$$

This expresses a relation which is posited to hold between certain physical magnitudes; in this case, a velocity v_i and some other variables on which it depends. Mathematically, we treat v_i as a dependent variable, and the x_k as independent variables. But mathematically, there is no reason to single out v_i in this way; mathematically, the relation is best expressed in a completely symmetric form:

$$\Phi(x_1, \ldots, x_n, v_i) = 0 \ . \tag{5}$$

Singling out the variable v_i explicitly as a dependent variable is tantamount to <u>superimposing</u> something on the purely mathematical relation (5). It is with that <u>something superimposed</u> that we shall be concerned.

Now the arguments of the relation (5) are formal analogs of primary precepts. More precisely, they are labels for the readings of appropriate <u>meters</u> (cf. Rosen 1978). The relation (5) says nothing about where the meters are, or when they are read; such details are entirely abstracted out of the mathematical relation. Accordingly, this single mathematical relation is subject to diverse <u>interpretations</u>. For instance, it may refer to n+1 different meters at the same point in space, read at a common instant. Or it may refer to a single meter at a fixed point in space, read at n+1 in- stants. Or it may refer to a meter at n+1 instants of space, which are read at up to n+1 different instants. Thus, the same relation may describe what is happening here and now, or here and there, or now and then.

To distinquish between these physically distinct situations, which are all represented formally by the same relation, we must impose further <u>structure</u> on the relation. That further structure amounts to a partition or classification of the arguments of the relation into distinct classes, and this in turn generates an <u>interpretation</u> of the relation, which is essential in conveying physical meaning. We do this tacitly all the time; such an act of interpretation or classification is why we write dynamics in the form (1) instead of the form (5). But <u>tacit</u> is not good enough for our purposes; we must be <u>explicit</u>.

We have seen in the preceding section that the concept of "information" can be got by looking at the effect of a variation in one meter reading on the reading of another meter. In that discussion, we <u>tacitly</u> supposed that all the meters in question were located at a common point, and read at a common instant. To be completely general, however, we must also open ourselves to other interpretations; we must consider the effect of a change <u>here and now</u> on what happens <u>there and later</u>. But this is <u>causality</u>, and this is why notions of "information", in its most general sense, find their most natural expressions in causal language.

Now causal language has itself long since completely disappeared from physics. Such perceptive students of science as Bertrand Russell have asserted (1912) that this is entirely a good and progressive thing. I am proposing, in a nutshell, that Russell was badly mistaken; that the disappearance of causal language was, in fact, a consequence of (a) the utilization of a mathematical language from which its essence had been abstracted, and (b) a <u>tacit</u> restriction to a certain conventional interpretation of this mathematical language. The result has been not a gain, but a loss; considerations of "information" force us out of these arbitrary limitations and conventions, to a richer (i.e., more structured) language in which what we all understand as theoretical physics is revealed as a limited

and extremely circumscribed special case. We shall say more about
this as we proceed.

IV. ON DYNAMICS, STATES AND CAUSES

As was pointed out in the preceding section, the tacit partition
of arguments appearing in a mathematical relation describing a
physical system has always been an essential feature in interpreting
the relation. This partitioning has many important corollaries,
even in the simplest case, where the arguments of the relation are
interpreted as meter readings pertaining to the same point in space
and the same instant in time. We have discussed this case at great
length elsewhere (e.g. Rosen 1978a, 1978b) and need not repeat it
here, except to mention that these partitions underlie all arguments
pertaining to symmetry and similarity in all systems, biological
and physical; we refer the reader to the cited works for details.

One conclusion of that work is that every mathematical relation
describing a physical system, (e.g. like (5) above), can be expressed
as a mapping

$$\Phi_g \ : \ X \longrightarrow Y \tag{6}$$

where the subscript g pertains to the genome of the system, X is the
space of environments, and Y is the space of system phenotypes. In-
tuitively, the genomic variables are those which convey specific
identity on the system in question; the space X consists of those
variables which can be set directly by an experimenter, and Y are
the remaining variables, whose values are determined from (6) once
identity and environment are specified.

As it stands, such partitions are devoid of either temporal or
causal attributes. However, there is a special subclass of such
relations which can be converted to causal form. This is the case
in which a phenotype $y \in Y$ in (6) is the velocity vector for its
associated environment $x \in X$; or more generally, of a projection Px
of x on some subspace of X. Then we can rewrite our equation (6)

in the form

$$y = d(Px)/dt = \phi_g(Px, (I-P)x) \tag{7}$$

We notice that in this case the environment vector $x \in X$ is further
partitioned into two parts: a part Px, whose velocity or time rate
of change is given explicitly by (7), and a part $(I-P)x$, whose
velocities are not specified. We can thus rewrite (7) in a familiar
form:

$$v = dz/dt = \phi_g(z, u) \tag{8}$$

where we have put $z = Px$, $u = (I-P)x$, $v = y$.

The quantity z in (8) is generally taken as the starting point
for theoretical science; it is usually called <u>state</u> or <u>state vector</u>
of the system. Indeed, the equation (8) is nothing other than the
conventional representation of a (generally nonautonomous) dynamical
system, with z as state vector and u as "control vector." But it is
crucial to realize that in the approach we have given above, the
concept of "state" has as yet no deep intrinsic significance; it is
merely a word which characterizes those observables which happen to
appear as arguments in a mathematical relation which also contains
their velocities. That is all.

Of course, the decisive feature of a relation like (8) is that
it implicitly involves the notion of time (or more precisely, of
the time differential dt) and may be rewritten in an <u>integrated</u>
form:

$$z(t) = \int_0^t \phi_g(z, u(\tau)) \, d\tau + z(0) \quad . \tag{9}$$

Now this integrated form is, like the ones we have considered
previously, a mathematical relation between meter readings. But
unlike those, it is subject to an entirely different <u>interpretation</u>.
Namely, it relates meter readings at one instant of time to readings

pertaining to different instants. With this interpretation comes
a new idea; that of _prediction_; the state variables z in (8) are
precisely those whose values can be predicted via the integrated
form (9). And this is also why the notion of state has been so
intimately associated with the notion of cause; an association which
we shall briefly explore.

In a nutshell, if we call the meter readings z(t) at the instant
t an _effect_, then we can say that:

1. The "initial state" z(0) is the _material cause_ of the effect
    z(t);

2. The genome vector g is the _formal cause_ of the effect z(t);

3. The operator $\int_0^t \Phi_g(..., u(\tau))d\tau$ is the _efficient cause_
    of the effect z(t).

We note in passing that there is nothing regarding _final cause_
visible in the above picture. This is, in fact, the fundamental
reason why final causes have been excluded from science. This
exclusion, based as it is upon what we have seen is extremely re-
strictive and limited (albeit influential) language, and even upon
a single tacit interpretation of that languge, may be quite
unwarranted. We have elsewhere argued at length (e.g. Rosen 1978c)
that physical systems may admit descriptions in which final causes
are perfectly definable, and indeed that such systems (usually called
anticipatory) may play a vital role in understanding organic behavior. But we need not pause to discuss these matters here.

Likewise, we shall not attempt to explore in this short space
the mathematical and epistemological consequences attendant upon the
transition from a picture based on parameterized families of
mappings from environments to phenotypes of the form (6), to a
different one based on the notion of a "state space" Z and a family
of vector fields described by (8) or (9). A fuller treatment is in
preparation. It is sufficient here to note that we have developed
a circle of ideas, as yet still very special, in which we talk

coherently about classifications of observables, about dynamics, and about causes. And still waiting in the wings, almost ready to be introduced into this already pregnant mixture, are the notions of activation and inhibition which carry with them the seeds of "information." Before doing this, however, we need to talk briefly about the notion of force.

V. ON FORCE AND CONSTRAINT

To keep our discussion simple, we shall confine our remarks to the confines of classical mechanics, the arena in which the concept of force was first coherently defined and applied.

In its most elementary form, classical mechanics may be subsumed into the theory of general dynamical systems like (1) or (8). Indeed, the modern theory of dynamical systems was regarded by its founders (e.g. Poincare, Birkhoff) as a direct generalization of mechanics. Thus, in (8), we might call the elements of Z phases instead of states; utilizing an appropriate imposed symplectic structure on Z, the relation (8) which expresses rate of change of phase in terms of phase can be related to accelerations and hence to forces. In this way, we can regard a force as specified by an equation of state which associates tangent vectors (phenotypes) to phases (environments) and re-interpret the discussion of the preceeding section accordingly. This can certainly be done, but it would obscure certain important matters, particularly those involving constraints.

We recall that in mechanics it is the configuration space which plays a distinguished role; our real interest is in how configurations change in time, and in a sense the phase space is a subsidiary construct which allows us to follow configurations conveniently. If we start from a configuration space Y, then a phase is a pair (y, v), where $y \in Y$, and $v \in T(y)$; i.e., v is a vector in the tangent space $T(y)$ at y.

In analytical mechanics, a constraint is simply a relation

$$\rho(y, v) = 0 \tag{9}$$

imposed on the phase space.   If the relation is independent of v,
the constraint is called holonomic; otherwise, nonholonomic.
Geometrically, the effect of a holonomic constraint is to reduce the
dimension of the phase space by 2; we lose one dimension of configuration, and hence also one dimension of velocity.   A non-holonomic
constraint, on the other hand, only reduces the dimension of the
phase space by one; we lose no configurational dimensions, but we
do lose a dimension of velocity; i.e., a degree of freedom.   If there
are r independent nonholonomic constraints, there are only n-r
degrees of freedom.   The effect of nonholonomic constraints, then,
is to restrict our choice of initial velocities, but not of initial
configurations.   They also serve to restrict the kinds of forces
which may be impressed on the system; i.e., to restrict the directions in which a tangent vector to a phase may point.

        Now let us notice that the constraint relation (9) is simply a
mathematical relation between observables y and some of their rates
of change.   But we have seen this situation before, and we can apply
to it the same argument which led to (7) above.   In particular, let
us write

$$y = y_1 + y_2; \quad v = v_1 + v_2$$

in such a way that

$$v_1 = dy_1/dt \quad .$$

Then the constraint relation can be written as

$$v_1 = dy_1/dt = \Psi(y_1; y_2, v_2) \quad . \tag{10}$$

This is exactly in the form (8), where y_1 is now a "state vector"

for a nonautonomous dynamical system, and the pair (y_2, v_2) is
the "control".

If there are r independent nonholonomic constraints, and hence
n-r degrees of freedom the dimension of the state space of this
nonautonomous dynamical system is n-r.   Therefore, if we impose
the maximum number of nonholonomic constraints, there will be <u>no
degrees of freedom left</u>; the dynamical system (1) will become <u>auto-
nomous</u>, and its state space will be identical with the original
<u>configuration</u> space.

These observations show us a new way to relate dynamical systems
to mechanics; one completely different from the one with which we
started.   Instead of regarding general dynamical systems as generalizations of mechanics, where the manifold of states generalizes phase
space and a vector field on the manifold of states generalizes
impressed force, we have a picture in which dynamical systems are
extremely special cases of mechanics.   The manifold of states is now
identified with configuration space rather than phase space; the
system is maximally constrained, and the dynamical equations arise
from forces of constraint rather than impressed forces.   In this
situation, the dynamics (10) is <u>invariant for all impressed forces</u>
compatible with the constraints; the only role of the impressed
forces is to start the system moving.

What we have indicated above comprise two entirely different
views of the relation between dynamical systems and the forces responsible for motions of mechanical systems.   The traditional picture
treats dynamical behavior as arising from (10), on the other hand,
regards it as arising primarily from constraining force.   There are,
thus, major epistemological differences in the causal interpretation
of the integrated forms of these situations, and hence, in discussing
the role of "information" in generating behavior.   In particular,
to say that "life is simply a matter of constraints" may have a far
deeper import than has traditionally been imagined.

VI.   INFORMATION, CAUSE AND EFFECT

We have come quite a long way in a relatively short space, and
perhaps, it would be well to pause here to review the salient points
before we proceed.

We have taken the basic position that natural science deals with
sets of meter readings, and with relations between them. These
readings may be taken from any kind of meter, located at any point
in space, and read at any instant of time. Traditionally, these
meter readings have been regarded as mere numbers, and the relations
between them are described purely as mathematical functions. But
as we have seen, this view ignores or abstracts certain crucial
properties of the actual situation, and this creates a corresponding
ambiguity of interpretation of these mathematical functions. The
ambiguity is conventionally ignored, because we tacitly supplement
the purely mathematical aspects with a kind of canonical interpretation. But this canonical interpretation is extremely limited, and
excludes much that is of decisive import to the physicist and the
biologist alike.

At the level of relations between meter readings, concepts like
state and cause have as yet no meaning. As we have seen, the concept
of state arises only when we have mathematical relations which involve observables and their rates of change. These observables are
then segregated out as state variables, and the mathematical
relations with which we began are then interpreted as equations of
motion. These can then be integrated to give new relations between
meter readings, and it is these relations which can be directly
interpreted in causal terms.

The interpretation of our original relations as equations of
motion is itself a special case of a wider situation, in which the
arguments of a relation between meter readings can be segregated into
genome. environment and phenotype. As we recall, the genomic observables convey identity, and serve as coordinates which identify
a particular mapping within a space of such; the environmental

circumstances; the phenotypic observables constitute the range of
this mapping, and are determined once genome and environment are.
In the special case of equations of motion, the environmental observables further partition into state variables and "forcings",
and the concomitant casual concepts arise in this picture of themselves.

In this context, we can finally re-introduce the simple ideas
of activation and inhibition with which we started. As we recall,
these ideas were developed entirely within the context of autonomous
dynamical systems, and we can now see just how special a situation
this is. Nevertheless, the ideas themselves can be directly extra-
polated to the wider context, at least in a formal way. What we
want to do is to measure "information" in terms of physical
magnitudes, utilizing the full circle of ideas introduced above.
Accordingly, by analogy with the activation-inhibition networks,
we would want to begin with the quantitites which can be expressed
symbolically as

$$\frac{\partial}{\partial (\text{cause})} \left(\frac{d}{dt} (\text{effect}) \right) \, .$$

These expressions are the crux of the relationship between information, dynamics and cause; whenever the latter two are meaningful,
so then is the first.

We can see at once from these expressions that there are (at
least) three different kinds of causes:

1. Genetic Information

As we have seen above, genomic observables in a physical relation are associated with formal cause. Therefore, what we may call
"genetic information" must be associated with the quantities

$$\frac{\partial}{\partial (\text{formal cause})} \left(\frac{d}{dt} (\text{effect}) \right) \, .$$

2. Somatic Information

By "somatic information" we shall understand that information associated with <u>material</u> <u>cause</u>. Accordingly, this kind of information is determined by the quantities

$$\frac{\partial}{\partial(\text{material cause})} \left(\frac{d}{dt}(\text{effect})\right) .$$

3. External Information

By this we shall understand the information related to <u>efficient</u> <u>cause</u>. Once the genome is determined, efficient cause is specified by those environmental magnitudes which are not state variables; i.e., by what is conventionally called "inputs" or "forcings" imposed upon a system. Under these circumstances, external information is associated with the quantities

$$\frac{\partial}{\partial(\text{efficient cause})} \left(\frac{d}{dt}(\text{effect})\right) .$$

Before going further, several features of this situation must be stressed. First, the three different kinds of information we have identified are, in fact, <u>all</u> <u>different</u>; different in mathematical expression, different in meaning, and different in interpretation. They cannot be interchanged or confused. Therefore, "information" is not simply one thing; it is, perhaps, the attempt to so regard it which is responsible for the extreme equivocation associated with this term. Second: the concepts we have introduced depend upon the prior specification of relations between meter readings, and are only meaningful relative to such a prior specification. Thus, "information" is not intrinsic, but relative to a mode of description. It appears to change as we enlarge, contract or otherwise modify our mode of description. As a matter of fact, we can study these changes in information as a function of description in the same manner that we study information itself, but we cannot enlarge on this matter

here. Third, the "information" we have been considering is in its
several forms of a semantic kind; not of a syntactic one. It is,
thus, I believe, more closely akin to what we heuristically mean by
"information" than is that of the Shannon theory, which enters at an
entirely different level.

We stated at the very outset that our purpose was to explore
the possibility that the language of theoretical physics was too
special for biology, rather than too general. We believe we have
made this possibility plausible, by showing that this traditional
language deals with only a very restricted class of relations between
meter readings; namely, those which pertain to meters pertaining to
a common point space and a common instant of time. A certain subclass of these (the equations of motion) allow us to formulate a
notion of state, and as we have seen, these lead to a new class of
relations in which classical notions of causality can be defined.
We used these notions in an essential way to show that informational
concepts are meaningful throughout the class of systems with which
physics conventionally deals.

We are now going to claim even more. Namely, we will argue that
our basic paradigm for characterizing "information" is in principle
valid, not just for the special subclasses of relations with which
physics has conventionally dealt, but throughout the entire class of
such relations, with which physics ought to deal, but usually does not.

Specifically, given any relation

$$\Phi(x_1, \ \cdot \ \cdot \ \cdot \ , \ x_n) \ = \ 0,$$

where the arguments x_i of this relation can refer to meter readings
at any point of space and any instant of time, traditional concepts
of state, cause and dynamics are simply not available. Yet, we can
still imagine a meaning which can be given to the expressions

$$\frac{\partial}{\partial x_j} \left(\frac{dx_i}{dt} \right) \ .$$

This meaning will not be a traditional one in general; i.e., one
which is immediately expressible in the language of traditional
calculus or its generalizations; but it can be done. Indeed, I would
argue that it _must_ be done if we are ever to have a truly general
theory of physics. When it is done, we can use these quantities,
which pertain to information, to work backward towards corresponding
generalizations of notions like cause. Specifically, if such an
expression should happen to vanish identically, this may be interpreted as saying that there is no causal connection between the
observable x_j and the observable x_i, at the positions and instants
at which these observables are evaluated. It remains to be seen
whether such generalized notions of causal connection can be further
partitioned into the traditional classes of formal, material,
efficient, and final causes. Indeed, we cannot say much at this
stage about how the physics of generalized relations will look. One
provocative observation, however, can already be made; namely, the
general asymmetry of the cause-effect relation will necessarily
manifest itself in the extreme noncommutativity of the resulting
formalism. Whether this noncommutativity is related to that already
so vividly manifested in quantum mechanics, it is too early to tell;
but the possibility provides a certain amount of food for thought.

VII. AN APPLICATION: COMPUTATION AND CONSTRUCTION

One of the most vivid features of organisms is that they propa-
gate; they reproduce. Ultimately, the reproduction of any organism
involves a notion of _self-reproduction_. Clearly, any general theory
of biology must come to grips with self-reproduction; indeed, it is
often felt that a theory of self-reproduction would provide a sufficient basis for everything else in theoretical biology.

Perhaps the most influential suggestion made in this direction
was that of the mathematician von Neumann (cf. Burks 1966). The
crux of von Neumann's argument for a "self-reproducing automaton"
is the postulation of the existence of a _universal_ _constructor_;
a machine which can build any other machine when provided with an

appropriate description (e.g. a blueprint). Von Neumann justified
this postulation on the basis of Turing's (1936) proof that there
exists a universal computer; i.e., a computer which will imitate the
computation of any special-purpose device when provided with a
description of it. The relation claimed by von Neumann between
construction and computation is essentially one between physics
and "information", and relies on the idea that a "machine" is
simultaneously a physical device to be built, and a logical device
which can execute a program and process information.

We will argue here that this argument of von Neumann is not
valid, and that the concept of a machine as a physical device has
nothing to do with the "mathematical machines" of which Turing spoke.
In effect, we shall argue that the utilization of a physical device
for purposes of computation involves aspects of efficient cause,
while construction involves material cause. Thus, arguments from
computation, including Godel's celebrated incompleteness theorem
(Godel, 1931), place no direct restrictions or constraints on
physical law.

Let us sketch the argument, at least within a single suitable
context. To that end, let us recall the relation (9) above which
will suffice for our purposes:

$$z(t) = \int_0^t \Phi_g(z, u(\tau)) \, d\tau + z(0) \quad .$$

In this relation, let us further identify:

    genome g = program (that which conveys identity);

    forcing $u(\tau)$ = input data;

    initial state $z(0)$ = starting state;

    mapping $u(\tau) \longmapsto z(t)$ = computation;

    mapping $z(0) \longmapsto z(t)$ = construction.

In this picture, specifying a genome g is tantamount to specifying
a particular (special-purpose) computer. The existence of a universal (i.e., general-purpose) computer may be expressed simply as

follows: there exists a particular genome g_o such that, for any
other genome g and input $u(\tau)$, we can write

$$\int_0^t \Phi_g(z, u(\tau)) \, d\tau = \int_{-T}^t \Phi_{g_o}(z, \bar{u}(\tau)) \, d\tau$$

Here $\bar{u}(\tau)$ has the property that

$$\bar{u}(\tau) = u(\tau), \quad 0 \le \tau \le t$$

$$\bar{u}(\tau), \quad -T \le \tau \le 0, \quad \text{"describes" the genome g.}$$

Thus, the universal machine imitates any special machine when an
input $u(\tau)$ to the special machine is prefaced by an input which
identifies or labels the special one. The salient point here is that
computation establishes a relation between efficient cause and output. In general, realizing such a relation in dynamical terms
requires the establishment of a superstructure of material causality
which the input can control. To use the terminology of H.H. Pattee:
computation is governed by rules which must be implemented. It
should be noted that a computer, in this sense, is closely related
to the notion of a measuring device or meter; what is measured is
also the efficient cause of the measurement.

Construction, on the other hand, establishes a relation between
material cause and output. A constructor is determined by fixing
both a genome g and a specific $u(\tau)$; this last must now be regarded,
not as as input to be processed, but as a set of "internal coordinates " of the constructor. (Previously, of course, the "state
variables" z(t) were treated as internal coordinates of the computer).
Further, it is the initial state $z(\emptyset)$ which is now regarded as variable (i.e., which now plays the role of "input" to the constructor);
previously, it was regarded as the fixed "initial starting-state"
of the computer.

From this, it is clear that the existence of a universal computer has nothing to do with the existence of a universal constructor;
indeed, in general, there is no such thing as a universal constructor.
We see, further, the vast distinction between a "mathematical
machine" and a physical device; the former, as we have noted, must
be interpreted as a relation between efficient cause and effect,
while the latter integrally involves a relation between material
cause and effect. Both relations look alike when considered from a
purely formal, mathematical point of view, but that is because the
mathematics has stripped away precisely the underlying structure
which we require to distinguish them physically.

We thus return to the point from which we started; from the
contention that the langugae of theoretical physics has been exces-
sively impoverished by its relentless neglect of organisms; so much
so that it has perhaps lost sight of its own true dimensions. In
establishing the true physical basis of organic phenomena, it is not
biology which will be swallowed up, as the reductionists believe; it
is, rather, physics which will be transformed, perhaps out of all
recognition. Indeed, the few ideas we have tried to sketch above
are doubtless only a small part of a much larger story, which as
yet has been but barely glimpsed.

REFERENCES

1. Burks A (ed). Theory of Self-Reproducing Automata. Urbana:
    University of Illinois Press, 1966.

2. Godel, K. "Uber formal unentscheidbare Sätze der Principia
    Mathematica and verwandter Systeme". Monatshefte für Mathematik
    und Physik 1931; 38:173-98.

3. Higgins, J. "The Theory of Oscillating Reactions". J. Ind.
    & Eng. Chem. 1967; 59:18-62.

4. Rosen, R. Fundamentals of Measurement and the Representation
    of Natural Systems. New York: Elsevier, 1978a.

5. Rosen, R. "Dynamical Similarity and the Theory of Biological
    Transformations". Bull. Math. Biol. 1978b; 40: 549-79.

6. Rosen, R. "Feedforwards and Global System Failure: A
    General Mechanism for Senescence". J. Theor. Biol. 1978c;
    74: 579-9Ø.

7. Rosen, R. "Some Comments on Activation and Inhibition".
    Bull. Math. Biol. 1979; 41: 427-45.

8. Russell, B. "On the Notion of Cause". 1912 Reprinted in
    Mysticism and Logic. New York: Norton, 1929.

9. Turing, AM. "On Computable Numbers". Proc. London Math.
    Soc. 1936; 42: 23Ø-65.

INFORMATION FLOW AND COMPLEXITY IN LARGE-SCALE METABOLIC SYSTEMS

Michael Kohn* and Samuel Bedrosian

University of Pennsylvania
Philadelphia, PA 19104
*Presented by

ABSTRACT

A graph-theoretical model of multienzyme networks is described.
Formal operations on the graph permit identification of the important
paths for information flow in the system and the enzymes which control the metabolic activity of the network. Simplification of the
graph facilitates the calculation of a normalized complexity index.

A real 12-enzyme system was only 5.6% as complex as a randomly
connected enzymic system, suggesting that metabolic pathways have
evolved into efficient feedback systems. The reason for this efficiency may be the tendency of enzymes to form function clusters;
enzymes which bind the same metabolites tend to communicate much more
strongly among themselves than with the rest of the network.

The processes of life are mediated by enzymes, proteins that
bind reactant molecules and hold them in the appropriate geometrical
configuration for a chemical reaction to occur. By providing general
acid-base catalysis, an enzyme lowers the activation energy of the
reaction, thereby increasing its rate. A metabolic pathway comprises
a sequence of enzyme-catalyzed reactions that converts a (usually
exogenous) chemical species into some end product. The network of

55

enzymic pathways in a cell either oxidizes fuel to liberate energy
necessary for the cell to perform its functions or converts "building
blocks" (e.g., simple sugars, amino acids) into the structures that
maintain cellular integrity.

The rates of most enzymatic reactions are under exquisite control. In addition to substrate (i.e., reactant) molecules, enzymes
bind molecules which do not participate in the reaction but modify
the enzyme's ability to bind reactants or catalyze the reaction.
Thus, these modifiers can be activators or inhibitors of the affected
process. An earlier graph model of metabolic systems (Kohn &
Letzkus, 1983) has been altered to make it more convenient for analysis by the usual graph-theoretical algorithms. In this model
metabolites (either substrates or modifiers) are represented by
circular nodes, enzymes by square nodes, and the relationships among
them by triangular nodes. The circles and squares are collectively
called chemnodes, and the triangles are called relnodes. A relnode
containing a + or - sign denotes an activator or inhibitor relationship, respectively, and an empty triangle denotes a substrate
relationship.

The nodes are linked by directed arcs as shown in Fig. 1; the
apex of the relnode points in the forward reaction direction.
Metabolite nodes are connected to enzyme nodes by a sequence of
relnodes. In Fig. 1, enzyme A binds metabolite 1 as a substrate and
releases metabolite 2 as the reaction product. If a modifier relnode
points to an enzyme node, the modifier affects the enzymes' catalytic
capacity. Inhibition by metabolite 2 in Fig. 1 is an example of
this. If a modifier relnode points to another relnode, the modifier
affects the ease of binding of the metabolite that is the immediate
ancestor of the target relnode. The binding of metabolite 3 by
enzyme A inhibits the binding of the substrate metabolite 1. We
note in particular that synergistic effects are possible. Thus, the
binding of metabolite 4 to enzyme A activates the binding of metabolite 3 as an inhibitor. That is, metabolite 3 is more inhibitory

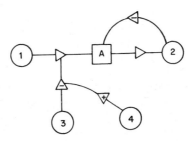

Fig. 1. Graph model of an isolated enzyme. Chemicals 1 and 2 are
        substrates of enzyme A, chemical 2 inhibits the catalysis by
        A, and chemical 3 inhibits the binding of chemical 1.   Chemi-
        cal 4 activates the enzyme for binding chemical 3, thus making
        3 more inhibitory than it would be in the absence of 4.

in the presence of large amounts of metabolite 4 than it is in the
presence of small amounts of metabolite 4.

     With each chemnode i is associated a number, c_i, the concentra-
tion of chemical species i. With each relnode is associated a pair
of numbers, K_{ij}, the binding strength of enzyme j for metabolite i,
and n_{ij}, the cooperativity of the binding (if $n_{ij} \neq 1$, several molecules of metabolite i influence each other's binding). For each
enzyme j there is a saturation function that tells what fraction of
the enzyme's catalytic capacity is being utilized for the given
metabolite concentrations. The saturation function for enzyme j is
given by

$$Y_j = \frac{1 - Q_j/Keq_j}{1 + \sum\limits_{\substack{sub-\\strates, i}} (K_{ij}/c_i)^{n_{ij}}} \quad ,$$

where Q_j, the mass action ratio is given by

$$Q = \prod_{\substack{\text{prod-} \\ \text{ucts, } i}} c_i \Big/ \prod_{\substack{\text{react-} \\ \text{ants, } i}} c_i,$$

and the equilibrium constant is given by

$$Keq_j = \prod_{\substack{\text{prod-} \\ \text{ucts, } i}} K_{ij} \Big/ \prod_{\substack{\text{react-} \\ \text{ants, } i}} K_{ij}.$$

If modifiers affect the enzyme's catalytic capacity, the saturation function becomes

$$Y'_j = Y_j \Big/ \Big[1 + \sum_{\substack{\text{activa-} \\ \text{tors, } i}} (K_{ij}/c_i)^{n_{ij}} + \sum_{\substack{\text{inhibi-} \\ \text{tors, } i}} (c_i/K_{ij})^{n_{ij}}\Big].$$

If modifiers affect the binding strength of a metabolite, the *effective* binding strength of that metabolite is given by

$$K'_{ij} = K_{ij} \Big[1 + \sum_{\substack{\text{activa-} \\ \text{tors, } k}} (K_{kj}/c_k)^{n_{kj}} + \sum_{\substack{\text{inhibi-} \\ \text{tors, } k}} (c_k/K_{kj})^{n_{kj}}\Big].$$

Enzyme graphs are concatenated as in Fig. 2a to form the graph of the complete pathway. Such a graph shows the flow of information as well as the flow of mass. The flow of information among separated enzymes is called feedback, and its function is to regulate the metabolic rate to be just enough for the cell's needs. Our strategy for identifying feedback paths is as follows:

1. Delete an enzyme node and all relnodes that define the enzyme's saturation function. This leaves a gap in the graph.

2. Find gap-spanning relnodes, the key points in paths by which information flows around the gap. It must be possible to trace a path from the immediate ancestor of a candidate gap-spanning relnode backwards to one chemnode on the gap boundary and trace another path from the immediate

descendant of the candidate relnode forwards to another
gap-boundary node for the inclusion of the candidate in the
set of gap-spanning relnodes. These paths must not contain
modifier relnodes, and the candidate relnode must have a
nonzero binding strength.

3. Find the minimum number of gap-spanning relnodes such that
their deletion will sever all lines of information flow
around the gap.

4. Replace the deleted enzyme and repeat this procedure until
all enzymes have been considered. Total feedback is represented by the union of all gap-spanning relnode sets.

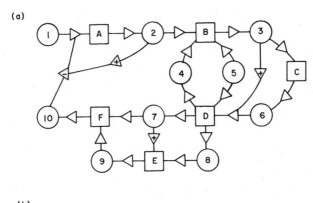

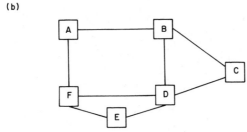

Fig. 2a.   Graph model of a six-enzyme system.
    b.   Reduced enzyme graph of Fig. 2a obtained by joining two
enzyme nodes if there is at least one path between them containing exactly one circular chemnode.

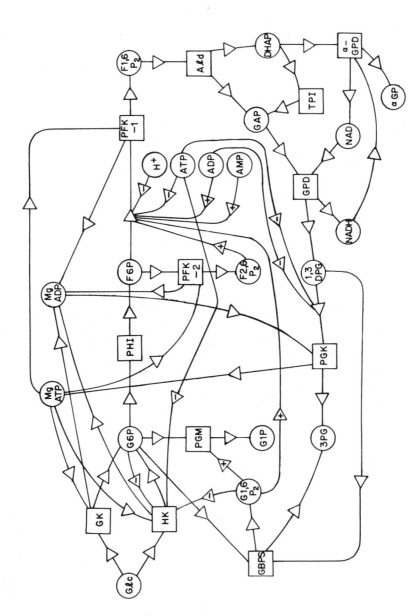

Fig. 3.   Graph model of glucose metabolism which includes the first
12 enzymes in the pathway.

The metabolites which are the immediate ancestors of the gap-
spanning relnodes are the feedback metabolites and the enzymes that
produce or consume those metabolites tend to control the activity
of the metabolic system. Enzymes in this group whose saturation
functions have the highest values are the most important in controlling the network. For example, deletion of enzyme B in Fig.
2a leaves a gap between metabolites 2 and 3. Feedback inhibition
by metabolite 10 is a gap-spanning relnode because a legal path can
be traced backwards from it to metabolite 3 and forward from it to
metabolite 2. Enzyme F, which produces metabolite 10, is a major
controller of the activity of this simple enzymic network. A more
realistic example is given by Fig. 3, which shows 70% of the pathway
for metabolism of glucose in liver or pancreatic islets. A complete
analysis of the graph for metabolism of glycogen (animal starch) has
been previously reported (Kohn & Letzkus, 1983).

Metabolic networks are generally considered to be complex
systems because of their large number of constituent chemical species
and the large number of interactions among these constituents. A
quantitative measure of the flow of information would be an objective
index of complexity. The number of paths for flow of information

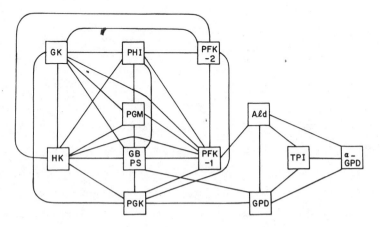

Fig. 4.   Reduced enzyme graph of Fig. 3.

among the enzymes is such a measure. As previously reported (Kohn
& Bedrosian, 1982), the number of such paths is given by the number
of spanning trees in a reduced graph containing only enzyme nodes.
The reduced graph is constructed by joining two enzyme nodes by an
undirected arc if there is at least one path between them in the
original graph that includes exactly one metabolite node (but an
unlimited number of relnodes). Fig. 2b gives the reduced enzyme
graph for the network in Fig. 2a, and Fig. 4 gives the reduced graph
for the network in Fig. 3.

In order to compare the complexity of systems with different
numbers of enzymes, we divide the number of spanning trees in the
reduced enzyme graph by the number expected if two enzymes were just
as likely to bind a common metabolite as not. The expected number
of spanning trees is

$$E(c) = n^{n-2} \frac{\binom{N/2}{n-1}}{\binom{N}{n-1}} \quad ,$$

where n is the number of enzymes and $N = n(n-1)/2$ is the number of
arcs in the complete graph K_n (Kohn & Bedrosian, 1982).

This approach is equivalent to asking a question about evolution. If the proteins in a metabolic pathway evolved to constitute
an efficient feedback system, the number of paths for information
flow should be small compared to the number expected if proteins
evolved so that they shared a metabolite pairwise with probability
1/2. If the proteins evolved to constitute a redundant feedback
system, the number of trees should be large compared to the expected
number. We can not decide *a priori* between these two possibilities.

There are 30 spanning trees in the graph of the 6-enzyme example
(Fig. 2b), and the expected number of trees is 9.06, giving a
normalized complexity index of 3.31. The reduced enzyme graph of
the 12-enzyme example (Fig. 4) has 3.3502×10^6 spanning trees and
an expected number of 5.9503×10^7, giving a normalized complexity

index of 0.056 (Kohn & Bedrosian, submitted MS). This result
suggests that a small portion of a metabolic system may seem somewhat complex relative to a randomly connected enzymic network, but
a multienzyme system that is more nearly the size of a real pathway
is far less complex relative to the same standard. Nature apparently
has evolved metabolic pathways to be highly efficient rather than
redundant feedback systems.

The reduced graph in Fig. 4 has 33 arcs, exactly the number
expected for a randomly connected 12-node graph, yet the number of
spanning trees is much smaller than the expected number for a
randomly connected 12-node graph. If the cause of this result can
be determined, it may indicate a possible evolutionary mechanism
for the origin of efficient feedback. It is known that the number
of spanning trees is maximized if all nodes are of the same degree
or differ at most by one. In Fig. 4 the degrees of the nodes vary
from three to eight. This is often indicative of the clustering of
nodes into relatively isolated groups. That this is the case here
is shown by Fig. 5, where the enzymes have been numbered for ease
of reference. Enzymes 1-8 form a group connected to another group
of enzymes 9-12 by only three arcs, the broken lines.

We define a (k,r)-cluster as a subgraph P of a graph G such that

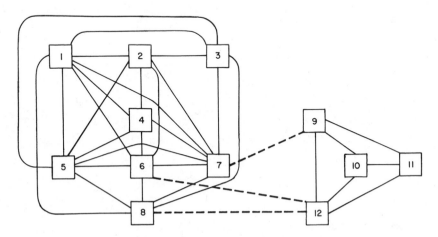

Fig. 5. Numbering assigned to the nodes in the graph of Fig. 4.
Broken lines indicate the clustering of the nodes into two groups.

the node degrees of P are $\geq$ k, the longest path length between any
pair of nodes in P is r < k, and P is not a subgraph of Q, another
subgraph of G, with the same properties as P. This definition is
midway between the single-link and complete-link cluster definitions
(Baker & Hubert, 1976). Preparata and Yeh (1974) have introduced
compatibility classes, generalizations of equivalence classes. A
compatability class can be represented as a maximally complete subgraph P (a clique, r = 1) of graph G. The complete cover of G is
the set of complete subgraphs of G such that every arc in G appears
at least once in the set. The fewer elements of the complete cover
that are necessary to account for all the arcs in a cluster the less
likely it is that the particular (k,r)-cluster would be as isolated
in a randomly connected graph as it is in the actual network.

The complete cover of the graph in Fig. 5 includes one K_6 (1,2,4,
5,6,7); three K_5's 1,2,3,5,7), (1,3,5,7,8), (1,5,6,7,8); one K_4 (9,10,
11,12); one K_3 (6,8,12); and one K_2 (7,9). The (5,2)-cluster (1-8)
is the union of the K_6 and the three K_5's, and the (3,1)-cluster
(9-12) is the K_4. So it would seem that the (3,1)-cluster (9-12)
is unexpectedly isolated, but that the (5,2)-cluster is not.

A direct comparison of the clustering properties of the reduced
enzyme graph with those expected for a randomly connected graph would
give the probability that a particular cluster could arise by chance.
Let the separation S_{ij} of nodes i and j be the number of arcs on the
shortest path between i and j. Table 1 gives this for the graph in
Fig. 5. Following Baker and Hubert (1975), we define the discrimination matrix D_{ij}^m for cluster m as the number of node pairs {p,q},
both nodes not in m, for which $S_{pq} < S_{ij}$. Table 2 gives these
matrices for our 12-enzyme example. The partition adequacy α_m (Baker
& Hubert, 1975) is the sum of the elements of D^m divided by the
largest value that sum could attain with any distribution of the
given number of arcs in m among the nodes of m. Clearly, α_m is
largest when no two nodes in m are joined by an arc (an anticlique).
So α_m varies from zero if m is a clique to one one if m is an
anticlique.

TABLE 1

UPPER TRIANGULAR

SEPARATION MATRIX

S_{ij} = Minimum path length (number of arcs)
between nodes i and j.

```
 1  1  1  1  1  1  1 |  2  3  3  2
    1  1  1  1  1  2 |  2  3  3  2
       2  1  2  1  1 |  2  3  3  2
          1  1  1  2 |  2  3  3  2
             1  1  1 |  2  3  3  2
                1  1 |  2  2  2  1
Node pairs in     1 |  1  2  2  2
(5,2)-cluster       |  2  2  2  1
                    |  -  -  -  -  -
                    |     1  1  1
                    |        1  1
                    | Node pairs  1
                    | in (3,1)-cluster
```

Baker and Hubert (1975) suggest as a measure of cluster validity

$$\gamma_m = 1 - 2\alpha_m \quad ,$$

which is normally distributed with a mean of about 1.5 $(\ln n)/n$ and standard deviation $1/n$, where n is the number of nodes in cluster m. (Strictly speaking, γ_m is normally distributed only if all distributions of the arcs among the nodes of m have distinct probabilities. This can always be assured, however, by adding a tiny random number to the probability of joining any two nodes in the random enzyme graph.)

TABLE 2

UPPER TRIANGULAR

DISCRIMINATION MATRIX

$$D_{ij}^{m} = \text{Number of node pairs, } \{p,q\} \text{ not both in cluster}$$
$$m, \text{ for which } S_{pq} < S_{ij}, \{i,j\} \text{ both in cluster } m.$$

(5,2)-Cluster

```
0  0  0  0  0  0  0
   0  0  0  0  0  9
      9  0  9  0  0
         0  0  0  9
            0  0  0
               0  0
                  0
```

(3,1)-Cluster

```
0  0  0
   0  0
      0
```

The (5,2)-cluster has a partition adequacy of 0.039 and a γ-statistic of 0.922; the (3,1)-cluster, being a complete subgraph, has $\alpha = 0$ and $\gamma = 1$. The expected γ values for the (5,2)- and (3,1)-clusters are 0.390 and 0.520, respectively, with respective standard deviations of 0.125 and 0.250. As both clusters have γ-statistics two or four standard deviations greater than the expected value, we believe that these clusters are unlikely to have arisen by chance. Comparison of Fig. 5 with the full graph of the system (Fig. 3) shows that enzymes which bind metabolites that are chemically similar tend to communicate much more among themselves then with the rest of the system. Metabolic networks apparently evolved efficient feedback mechanisms by functional coupling of enzymes that catalyze similar chemical reactions or are involved in the metabolism of species that are in the same chemical family.

ACKNOWLEDGEMENTS

We wish to thank Prof. Saul Gorn, Department of Computer and Information Science, University of Pennsylvanis for his helpful advice during the conduct of this work.

Supported in part by Public Health Service Grant HL 15622 and an extension of work started under Office of Naval Research Contract N00014-75-C-0768.

REFERENCES

Baker, F. B., and L. J. Hubert, Measuring the Power of Hierarchial Cluster Analysis. J. Am. Statist. Ass., 70, 31 (1975).

Baker, F. B., and L. J. Hubert, A Graph-theoretic Approach to Goodness-of- fit in Complete-link Hierarchical Clustering. J. Am. Statist. Ass., 71, 870 (1976).

Kohn, M. C., and S. D. Bedrosian, Complexity of Metabolic Networks. Proc. 35th Ann. Conf. Engng. Med. Biol., 233 (1982).

Kohn, M. C., and S. D. Bedrosian, Complexity of Large-Scale Metabolic Networks. Trans. IEEE, (AC, CAS and SMC), submitted for the joint Special Issue on Large Scale Systems.

Kohn, M. C., and W. J. Letzkus, A Graph-theoretical Analysis of Metabolic Regulation. J. Theor. Biol., 100 (1983), in press.

Preparata, F., and R. T. Yeh, Introduction to Discrete Structures, Addison-Wesley, Reading, MA, 1974.

PHYSICAL PRINCIPLES AND PROTEINOID EXPERIMENTS IN THE EMERGENCE OF LIFE

Sidney W. Fox

University of Miami

Coral Gables, FL   33124

Professor Wald's paper, to use an old cliche, is a hard act to follow. In a real sense, however, those few of us who are self-organizationalists working on the problem of the origin of life have necessarily long been following Professor Wald. Each time we make an advance in the laboratory, we are likely to find, as I have on several occasions, that George Wald (1954) had perceived and expressed the essential concept earlier.

The history of the idea that matter might organize itself into living beings goes back to Louis Pasteur in 1864 (Vallery-Radot 1922) who asked,

> "Can matter organize itself? In other words,
> are there beings that can come into the world
> without parents, without ancestors?"

It is to this question that Wald (1954) first gave a positive answer. At the time F.O. Schmitt (1956) was studying the self-organizing properties of a protein, collagen. When Schmitt allowed a solution of collagen to evaporate on the grid of an electron microscope, a featureless film (Fig. 1A) remained. If, however, collagen was precipitated from solution by pH adjustment, the result was microfibrils having periodic striation (Fig. 1B).

1A   1B   1B

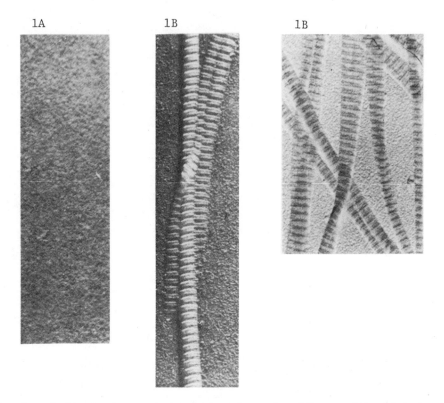

Fig. 1. A first cleancut demonstration of self-assembly (Schmitt,
         1956).
         A. Collagen evaporated onto a grid for electron micro-
             scopy.
         B. Two examples of selforganized microfibrils of collagen
             precipitated from solution.

From this work of Schmitt and associates, the idea of self-
organization ("self-assembly") has grown until it is now one of the
basic and pervasive tenets of physical biochemistry and physical
biology (Lehninger 1975 Chapter 36). In conceptualizing the origin
of life, Wald applied this principle with a statement,

"We have therefore a genuine basis for the
view that the molecules of our oceanic broth will
not only come together spontaneously to form
aggregates but in doing so will spontaneously
achieve various types and degrees of order. - - -
given the right molecules - - - they do a great
deal for themselves."

Part of Wald's description for the oringin of life was

amino acid $\rightleftharpoons$ protein $\longrightarrow$ aggregate   ;

"aggregates" became the first organisms. This sequence is close
to the first sequence of events that our experiments subsequently
revealed (Fox 1960a). The double-headed arrow undoubtedly signified
the recognized equilibrium between amino acids and proteins in
aqueous solution and that equilibrium rests close by amino acids
(Huffman 1942). We altered the equilibrium to one favorable to
polypeptides by driving water off from the reaction mixture (Fox
and Dose 1977). Such shift is entirely geological; e.g., evapora-
tion is constantly occurring on this planet. Once formed, the
thermal protein proves to be indefinitely stable in aqueous medium
(cf. Rohling 1970), as are the microspheres. From this fact, we have
extended the old thermodynamic theory to embrace newer concepts such
as stabilization by intramolecular reactions. Interactions are,
accordingly, stabilizing forces, i.e., they provide molecular trusses
in macromolecules. Thermal proteins seem to have been what Wald
referred to as the "right molecules."

The experimental demonstration of these steps, including self-
organization, was complete by 1960 (Fox 1960a) and the significance
of experimental self-organization was indicated in a title of Self-
organizing Phenomena and the First Life (Fox 1960b). More highly
theoretical treatments of self-organization are indicated in titles
of Selforganization of Matter and the Evolution of Biological
Macromolecules by Eigen (1971) and Self-organization in Nonequilibrium Systems by Nicholis and Prigogine (1977).

The concept of selforganization has been attacked (Yockey 1981) and defended (Fox and Matsuno 1983).

The aggregates obtained by thermal copolymerization (Fox and Dose 1977) of amino acids were initially found experimentally to be remarkably cell-like in appearance and, in later studies (Fox 1978), in function. A theory of formation of the precursor polymers was made possible through rejection of a number of popular assumptions (Fox and Dose 1977) and an awareness of the utility of trifunctional amino acids in copolymerization. Earlier awareness of the possibility for self-sequencing of amino acids (Fox 1956) was also crucial.

A modified form of the flowsheet of events that appears in some biochemistry textbooks (e.g. Florkin 1975, Lehninger 1975) is presented in Fig. 2. One may see in the second through fourth stages a sequence much like that which Wald presented earlier.

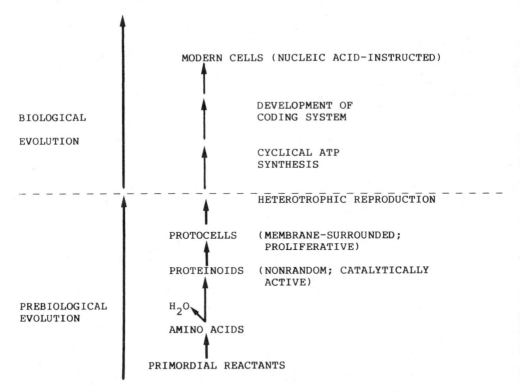

Fig. 2 Flowsheet of molecular evolution as deduced from results of proteinoid experiments.

The experimental aggregates, however, are found to be immediately cell-like structures (Fig. 3) arising in a single, simple, direct conversion from the noncellular precursor. The size and shape resemble those of coccoid bacteria; if anything, the units are seen to be more orderly than a colony of bacteria. One gm of polymer typically yields 10^{10} proteinoid microspheres. Their order appears to match the order in the precursor polymers, and is undoubtedly ascribable to that source.

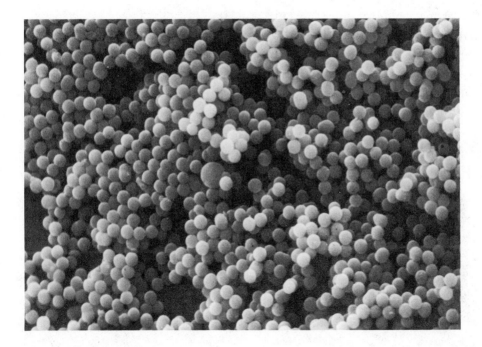

Fig. 3. Scanning electron micrograph of proteinoid microspheres (courtesy of Steven Brooke Studios). These laboratory protocells (approximately 1 μm in diameter) are quite uniform and extremely numerous. Studies of their functions suggest that they represent protolife from which modern life evolved. See Fig. 7.

EVIDENCE FOR NONRANDOM POLYMERS

The first kind of evidence for ordered polyamino acids was that of limited, i.e. nonrandom heterogeneity (Fox 1981). It is illustrated by the chromatogram of hemoproteinoid of Dose and Zaki (1971) (Fig. 4). For this proteinoid, Dose and Zaki heated the modern 20 common amino acids and a small amount of heme. The product was a hemoproteinoid containing tightly bound heme. It had enzymic (peroxidase) activity. Following a preliminary purification (dialysis), it proved to be remarkably homogeneous on discgel electrophoresis. Other thermal polyamino acids show similar limitation in heterogeneity. The products are thus nonrandom and since, in most cases, only amino acids are present at the start of reaction, the amino acids are self-ordering. The various kinds of evidence for self-ordering are numerous (Fox and Dose 1977, Fox 1980).

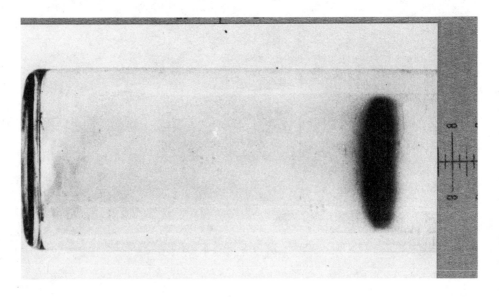

Fig. 4. Electropherogram of hemoproteinoid (Dose and Zaki 1971). The single bond indicates sharply limited heterogeneity.

A special kind of evidence of self-ordering is that which
indicates that the <u>sequences</u> of amino acids is extremely nonrandom
comparable to that observed through the modern coding mechanism.
In the most thoroughly studied demonstration, glutamic acid,
glycine, and tyrosine were heated to form a polymer. Following
solution in slightly alkaline medium (pH 7.5) the peptide product
was fractionated on Sephadex and then fractionated by paper chromatography (Nakashima, et. al. 1977). A number of minor peptides
were obtained (Fig. 5). Surprisingly, at first, these were all
discrete peptides and their amino acid compositions were stoichio-
metric. Many were pryoglutamylpolyglycyltyrosines, having various
numbers of glycine residue. The dominant fraction, however, was
an equimolar complex of two tripeptides:

       pyroglutamyglycyltryrosine

       pyroglytamyltyrosylglycine.

No other tripeptides were found. This result has been confirmed by
workers in Mainz (Hartman, et. al. 1981). In collaboration with
chemists in Frankfurt, Dose (Heinz, et. al. 1979) had already
learned that pigments of flavin and pterin constitution were formed
during the heating. The inclusion of these pigments as 1.7% of
weight of the polymer can explain that tripeptides were obtained as
a dominant fragment during the early fractionation due to fission
at pigment-peptide nodes. Dr. John Jungck calculated that the
amount of tripeptide recovered is 19.2 times (Nakashima, et. al.
1977) that anticipated on the basis of the random hypothesis (Eigen
1971 a,b).

The identities of peptides expected on the basis of the random
hypothesis vs. those found are listed in Table I. It can be seen
immediately that the actual result was highly nonrandom. Numerous
confirmations and extensions of the result of nonrandomness in the
products of heated amino acids have been accumulated (Fox 1980b,
1981). Selective interactions of amino acids or derivatives in
aqueous solution have also been reported (Calvin 1969, Cavadore
1971). The latter show less precision in self-sequencing; this is

Table I

Tyrosine-containing Tripeptides
Found vs. Those Expected on the
Basis of the Random Hypothesis

Expected from *Random* Polymerization		*Found* from *Nonrandom* Polymerization
αUαUY	YαUU	
αUγUY	YγUU	
γUαUY	YαUG	
γUγUY	YγUG	
αUGY	YαUY	
γUGY	YγUY	
αUYU	PαUY	
γUYU	PγUY	
αUYG	PGY	PGY
γUYG	PYU	
αUYY	PYG	PYG
γUYY	PYY	
GαUY	YGU	
GγUY	YGG	
GGY	YGY	
GYU	YYU	
GYG	YYG	
Gyy	YYY	

The dominant fraction obtained from the thermal copolymerization
of glutamic acid, glycine, and tyrosine proved to be an equimolar
complex of pyroglutamylglycyltyrosine and pyroglutamyltyrosyl-
glycine (Nakashima et al. 1977, Hartmann et al. 1981).

U = glutamic acid residue
Y = tyrosine residue
G = glycine residue
P = N-pyroglutamyl

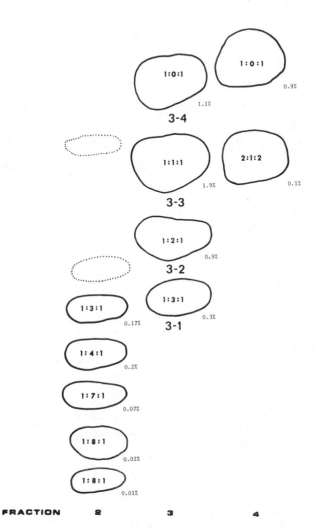

Fig. 5. Discrete peptide by paper chromatography of thermal product of glutamic acid, glycine, and tyrosine (Nakashima, et. al. 1977, Hartmann, et. al 1981). Dominant fraction is 3–3; it represents all tripeptides formed.

understood on the basis that the reactants are diluted by much water.

While the evidence for self-ordering of amino acids has been amassed by many methods in many laboratories (Fox 1980b), the truly striking feature is the high precision of the self-ordering process. It is as invariant as any result from repetitions of a reaction in the chemical laboratory, and is comparable to the precision of the coding mechanism.

The concept of nonrandomness is one that has had to contend with the random hypothesis, an hypothesis that is at the heart of the neo-Darwinian premises (Ho and Saunders, 1983) and which pervades some aspects of physics (Matsuno 1982). This concept likens amino acids to playing cards. This view is illustrated in Fig. 6. The drawn representations of molecules as either playing cards or inter-digitating shapes are outrageously inaccurate.

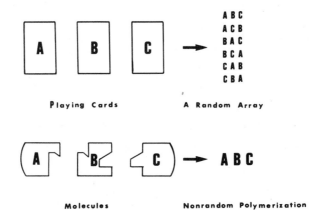

Fig. 6. In accord with game theory, shuffling and laying out of three card yields random array. Molecules, amino acid types, has each a unique shape, unlike playing cards. From three amino acid types, one or more tripeptide is preferred.

Much the greater outrage is perpetrated by the playing card image, since molecules do have at least different shapes. Evolution is thus stereomolecular. In its dependence upon shapes of molecules, evolution is, notably, susceptible to physical analysis.

SIGNIFICANCE OF SELF-ORDERING

One primary significance of the self-ordering was that it indirectly solved the old chicken-egg problem of which came first: protein or nucleic acid? In modern systems, each is needed for the other. The two requisite functions are ordering of amino acids in proteins, and catalysis of formation of bonds that occur in protein or polynucleotide. The self-ordering results indicate that amino acids could have provided the instructions for their own sequencing in the first proteins; neither DNA nor RNA were needed.

The view that DNA or RNA are primal in modern systems and that they are "self-replicating" is or has been prevalent. Some shift in this outlook has occurred, due especially to work by Kornberg on the DNA polymerases (Kornberg 1980). Kornberg points out that an error of omission in the 1953 Watson-Crick formulation of DNA replication was the failure of the authors to think in terms of the enzymes studied by the biochemist:

> "The suggestion was made in 1953 that A-,T-, G-, and C-containing precursors might orient themselves as base pairs with a DNA template and then be 'zippered' together without any enzyme action. However, to the biochemist it is implicit that all biosynthetic and degradative events are catalyzed by enzymes, making possible refinements of control and specificity, as well as rapid rates of reaction."

This emphasis on the primal function of informational proteins in the production of polynucleotides, which are also informational, is consistent with the emphasis provided by thermal proteins arising from self-instructed amino acids. While it conflicts with assumptions of Crick (1981) and Eigen, et. al., (1981), the experimentally

based view that a cell would have arisen before the polynucleotides
(cf Dyson, 1982) is even more in fundamental disagreement with
assumptions of Crick, Eigen, et al. Proteins-first and cells-first
are the only premises yet supported by a comprehensive experimental
model, or by any laboratory model, for that matter.

ANABOLIC AND CATABOLIC THERMAL PROTEINS

Consistent with the endogenously generated order in thermal
proteins, as seen in a number of laboratories (refs. in Fox 1980a),
these polymers have many catalytic activities of the kind that
compose cellular metabolism (Fox 1980a). Of primary interest to
our problem are the anabolic, or synthetic, activities. The various
catalytic activities, including the synthetic ones, one incorporated
into the cellular structures when the proteinoids are aggregated to
form them.

Such synthetic activities are found especially in the basic
proteinoids. Although the basic proteinoids do not readily form
spherules (Rohlfing 1975) as do the acidic proteinoids, they do so
in combination with acidic proteinoids. When they undergo such
coassembly the particles in suspension display the same synthetic
activites that the basic proteinoid alone exhibits in aqueous
solution.

Of special interest is the fact that the basic proteinoids,
in either solution or in suspended particles, catalyze also the
synthesis of the internucleotide bond, which is the backbone bond
of DNA and RNA. The study is at the stage at which we are investi-
gating how large are the cellular peptide and polynucleotides that
could initially arise in the way described.

Even so, what we now see is a cellular locale in which one
catalytic agent, basic thermal peptide, conceptually could have
served to make both polypeptides and polynucleotides in intimate
association (Fox 1982). This is then an ideal locale for the origin
of the genetic code and mechanism through direct interactions, even
though the processes may have been replaced by indirect reactions

involving t-RNA in later evolution (Dillon 1978).

Evidence for direct interactions between polynucleotides and polyamino acids has been accumulated; these interactions are stereo-chemically selective (Yuki and Fox 1969, Fox 1974, Lacey et. al. 1979). The selection can be codonic, anticodonic, or otherwise-depending upon conditions. While these experiments do not, of course, represent part of the actual coding or decoding process, they illustrate forces of molecular recognition that would plausibly have served. How these would have been permanently installed to a universal code, by *feedback fixation* has been suggested (Fox 1978).

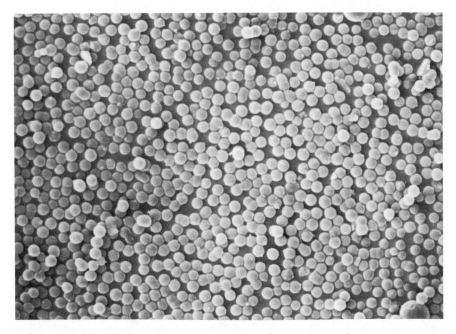

Fig. 7. Scanning electron micrograph of proteinoid microspheres of copoly (asp, gly, arg), quite basic. By Robert M. Syren. Average size 2.2 μm.

The proteinoid microsphere of the kind seen in the scanning electron micrograph of Fig. 7 differ from those of Fig. 3 by containing a substantial proportion of basic polymer, although the appearance is hardly distinquishable.

The inferred flowsheet for molecular evolution is indicated in Fig. 8. The possibility that a nontemplated protein synthesis

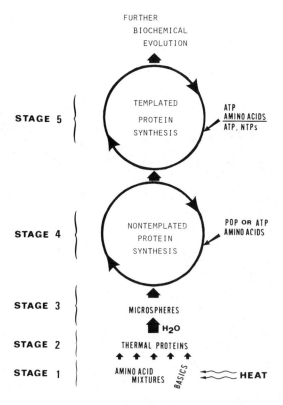

Fig. 8. Enlarged flowsheet for molecular evolution. (Thermal) peptides make other peptides (with ATP in water). Nontemplated protein synthesis probably preceded templated protein synthesis, which has been partially modelled experimentally.

evolved to a templated protein synthesis is acknowledged by Eigen and Schuster (1979); it is the possibility for which a demonstration of steps has been forthcoming.

THE MULTIACTIVE PROTOCELL

The numerous activities that have been catalogued in the proteinoid microsphere (Fox and Nakashima 1980, Przybylski et al. 1982) make it functionally and somewhat morphologically different from the inert structure visualized as a protocell (protobiont) by Oparin (1957) and others. Oparin's evaluations arose from his studies of coacervate droplets, which are intrinsically inert. The numerous activities of the proteinoid microspheres have, in fact, made possible the new evolutionary science of protobiology (Fox 1983).

The experiments have revealed that the proteinoid protocell was a many-faceted microstructure. Did it have all the main features that would have permitted it to evolve to a modern cell? While this question has been extensively analyzed, the main features for which modern analysts believe that primordia should be discernible are: (a) a molecular ordering mechanism, (b) metabolism, and (c) membrane. Heterotrophic growth and reproduction were shown early (Fox and Dose 1972). The protein would obviously have provided metabolic activities, while the history of endogeny stemming from the self-sequencing of amino acids would have answered the problem of ordering. The property of membrane is found in the proteinoid microsphere boundary, without any contained phospholipid. The kinds of evidence are numerous (Przybylski et al. 1982). To these are now added substantial conformance of this artificial to the constraints of the Hodgkin-Huxley equation derived from natural cells (Stratten 1982).

BIOELECTRIC ACTIVITY

Of the now numerous cellular properties catalogued in these microspheres (Fox and Nakashima 1980), the latest is membrane electricity, as seen in membrane potentials, oscillatory potentials,

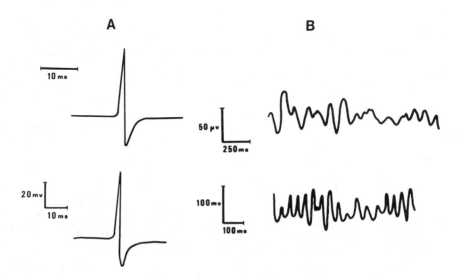

Fig. 9 (A) Tracings of single spiking discharge from squid giant axion (upper) and from proteinoid (2:3:1) microsphere. Upper tracing from Bures, Petrian, and Zachner (1967).

(B) Upper tracing of EEG from monkey in deep sleep (Jasper as copied in Woody, 1982). Lower tracing from microelectrode placed in dense suspension of proteinoid (2:2:1) microsphere in artificial seawater.

and spiking (Przybylski et al. 1982). This is the only protobiological function of many (Fox and Dose 1977, Fox 1980a) to be
discussed in these pages.

Typical results are seen in Figs. 9A and 9B.

This result and others like it are interpreted to signify that
the first cells on Earth already had, among numerous properties,
that which had long been thought of as a most advanced function:
electrical potentials. The need for special cells for electrical
activity, however, had some years earlier been repudiated, especially
with spiking having been observed in such lowly organisms as the
alga *Nitella* (Findlay 1959), and even in vesicles expressed from
Nitella (Takenaka 1971). Membranes slowly formed in the latter.

SUMMMARY

In partial summary, those physical principles quickly reviewed
in this presentation stem from a self-sequencing of amino acids.
Self-sequencing is a precise form of self-ordering and the self-
ordering of monomers is a seminal phenomenon in the type of self-
organization that yielded the first cell. This is a detailed
affirmative answer to Pasteur.

This phenomenon of self-organization including self-sequencing
resolved the old chicken-egg problem by indicating that ordered
proteins first arose by reaction of amino acids. *The instructions
were in the shapes of the amino acids; nucleic acids were not
necessary initially.* Such results do not disqualify as an alternative the early enzymeless replication of polynucleotides (Lewin
1982). Enzymeless "self-replication" of RNA has not been accom-
plished, in accord with the emphasis in the quotation of Kornberg
(1980). On the other hand, the protoenzymic synthesis of polynucleotides has been demonstrated (Fox 1981b) and seems to be a
process uniquely susceptible to providing evolutionary advantages.
This is comparable to the principle of cells-first, which has been
placed on a physical basis (Matsuno 1982), and to which principle
the protomechanism for coding is linked (Fox 1981b). Moreover,

the production of an array of protoenzymes, as we visualize for
proteinoids, would lead to a *general* metabolism, which must be
explained for some evolutionary stage anyhow.

In addition to the principle of self-sequencing wrapped within
self-organization, a second main lesson from these studies is that
the protocell already had within it the potentialities for further
evolution. That protocell arose in huge numbers with the greatest
of ease. The fact that a cell emerges easily would have provided
numerous evolutionary advantages (Fox 1976), e.g. the benefit of
a biochemical microcosm in which reactants are maintained in
proximity.

Examination of this simulated protocell has shown ordered
macromolecular composition, a barrier membrane, and metabolism
including the synthesis of peptide and internucleotide bonds of
protein and nucleic acid, respectively. In addition, the laboratory
protocell can grow (heterotrophically) and reproduce, and it
displays electrical excitability.

The functionalities by which the proteinoid microspheres differ
from the popular paradigm are numerous (Fox 1983). Outstanding
differences between the popular and proteinoid paradigms are the
self-ordering principles at the molecular level and the early
emergence of a cell. In these two principles, the experimental
results especially distinguish the proteinoid model from inferences
of experts such as Eigen and Schuster (1979), Nicolis and Prigogine
(1977), and Crick (1981).

Two principal findings seem to be (a) a protocell and (b)
excitability in the protocell. Our interest in the latter is
related to the origin of consciousness in much the same way as has
been expressed by Kuffler and Nicholls (1976):

> "the human brain consists of over 10,000
> million cells and many more connections that in
> their detail appear to defy comprehension. Such
> complexity is at times mistaken for randomness;

yet this is not so, and we can show that the brain
is constructed according to a highly ordered
design. - - - Studendts who become interested in the
nervous system almost always tell us that their
curiosity stems from a desire to understand perception,
consciousness, behavior, or other higher functions
of the brain. - - - they are frequently surprised that
we ourselves started with similar motivations - - -
and that we have retained those interests - - -."
Wald (1983) has recognized the problem of consciousness in its
relationship to matter, a relationship that was necessarily absent
from Pasteur's statement. Experiments now permit us to entertain
the possibility that when the right kind of matter, thermal protein,
organized itself the evolutionary beginnings of both life and mind
had emerged.

An early appeal for a physical study of the relationship
between matter and consciousness is that of Wigner (1961). The
reference in my title to the "emergence of life" rather than to the
usual "origin of life" is a reference to the view that one real
origin was the Big Bang; all other events are transformations from
the Big Bang (Fox 1980b). Full recognition of this view positions
the problems of emergence squarely in the domain of physics.

Acknowledgment. The research related in this paper has been aided
mainly by Grant NGE 10-007-008 of the National Aeronautics and
Space Administration. Contribution No. 357 of the Institute for
Molecular and Cellular Evolution. Thanks are expressed to Professor
Koichiro Matsuno for discussion of this paper, and to Professor
Aleksander Przybylski and Robert M. Syren for illustrations.

REFERENCES

Bures J Petran J and Zacher U 1967 Electrophysiological Methods
   in Biological Research Academic Press, New York 333.

Calvin M 1969 Chemical Evolution Oxford Univ. Press 169.

Cavadore J-C 1971 Polycondensation d' α-amino acides en milieu aqueux These Acad Montpellier.

Crick F 1981 Life Itself Simon and Schuster, New York.

Dillon L S 1978 The Genetic Mechanism and the Origin of Life Plenum Press, New York.

Dose K und Zaki L 1971 Hämoproteinoide mit peroxidatischer und katalischer aktivität Zeit Naturforsch 26b 144-148.

Dyson F 1982 A model for the origin of life J Mol Evol 18, 344-350.

Eigen M 1971a Selforganization of matter and the evolution of biological macromolecules Naturwissenschaften 58 465-523.

Eigen M 1971b Molecular self-organization and the early stages of evolution quart Rev Biophys 4 149-212.

Eigen M and Schuster P 1979 The Hypercycle Springer-Verlag, Heidelberg 83.

Eigen M Gardiner M Schuster P and Winkler-Oswatitsch R 1981 The origin of genetic information Sci American 244(4) 88-118.

Findlay G P 1959 Studies of action potentials in the vacuole and cytoplasm of Nitella Austr J Biol Sci 12 412-426.

Florkin M 1975 Ideas and experiments in the field of prebiological chemical evolution. In: M Florkin and E H Stotz, Eds Comprehensive Biochemistry Vol. 29 Part B Comparative Biochemistry, Molecular Evolution Elsevier, Amsterdam, 231-260.

Fox S W 1956 Evolution of protein molecules and thermal synthesis of biochemical substances Amer Scientist 44 347-359.

Fox S W 1960a How did life begin? Science 123 200-208.

Fox S W 1960 B Self-organizing phenomena and the first life Yearbk Soc Genl Syst Res 5 57-60.

Fox S W 1974 Origins of biological information and the genetic code Molec Cell Biochem 3 129-142.

Fox S W 1978 The origin and nature of protolife. In: W H Heidcamp, Ed, The Nature of Life University Park Press, Baltimore 23-92.

Fox S W 1980a Metabolic microspheres Naturwissenschaften 67 378–
    383.

Fox S W 1980b Life from an orderly Cosmos Naturwissenchaften 67
    576–581.

Fox S W 1981a Copolyamino acid fractionation and protobiochemistry
    J Chromatogr 215 115–120.

Fox S W 1981b Origins of the protein synthesis cycle Intl J Quantum
    Chem QBS8 441–454.

Fox S W 1983 Proteinoid experiments and evolutionary theory.   In:
    M-W Ho and P T Saunders, Eds. Beyond Neo-Darwinism Academic
    Press, London.

Fox S W and Dose K 1972 Molecular Evolution and the Origin of Life
    Freeman, San Francisco.

Fox S W and Dose K 1977 Molecular Evolution and the Origin of Life,
    rev. ed. Marcel Dekker, New York.

Fox S W and Matsuno K 1983 Self-organization of the protocell was
    a forward process J Theor Biol 00 000–000.

Fox and S W And Nakashima T 1980 The assembly and properties of
    protobiological structures: the beginnings of cellular peptide
    synthesis BioSystems 12 155–166.

Fox S W Nakashima T Przybylski A and Syren R M 1982 The updated
    experimental proteinoid model Intl J Quantum Chem QBS9 195–204.

Hartmann M Brand M C and Dose K 1981 Formation of specific amino
    acid sequences during thermal polymerization of amino acids
    BioSystems 13 141–147.

Heinz B Ried W and Dose K 1979 Thermal generation of pteridines
    and flavins from amino acid mixtures Angew Chem Intl Ed Engl
    18 478–483.

Ho M-W and Saunders P T 1983 Beyond Neo-Darwinism Academic Press,
    London.

Huffman H M 1942 Thermal data XV The heat of combustion and free
    energies of some compounds containing the peptide bond J Phys

Chem 46 885-891.

Kornberg A 1980 DNA Replication Freeman, San Francisco.

Kuffler S W and Nicholas J G 1976 From Neuron to Brain, Sinauer
    Associates, Sunderlan MA viii.

Lacey J C Jr Stephens D P and Fox S W 1979 Selective formation of
    microparticles by homopolyribonucleotides and proteinoids rich
    in individual amino acids BioSystems 11 9-17.

Lehninger A L 1975 Biochemistry Worth and Co., New York.

Lewin R 1982 RNA can be a catalyst Science 218 872-874.

Matsuno K 1982 A theoretical construction of protobiological synthesis:
    from amino acids to functional protocells Intl J Quantum Chem
    QBS9 181-193.

Nakashima T Jungck J R Fox S W Lederer E and Das B C 1977 A test
    for randomness in peptides isolated from a thermal polyamino acid
    Intl J Quantum Chem QBS4 65-72.

Nicolis G and Prigogine I 1977 Self-organization in Nonequilibrium
    Systems John Wiley and Sons, New York.

Oparin A I 1957 The Origin of Life on the Earth, Academic Press,
    New York.

Przybylski A Stratten W P Syren R M and Fox S W 1982 Membrane, action,
    and oscillatory potentials in simulated protocells, Naturwissen-
    schaften 69 561-563.

Rohlfing D L 1970 Catalytic activities of thermally prepared poly-
    α-amino acids: effect of aging Science 169 998-1000.

Rohlfing D L 1975 Coacervate-like microspheres from lysine-rich
    proteinoid Origins Life 6 203-209.

Schmitt F O 1956 Macromolecular interaction patterns in biological
    systems Proc Amer Philos Soc 100 476-486.

Stratten W P 1982 Comparison of electrical properties of nerve
    membrane to those of proteinoid-lecithin spheres Abstracts Soc
    Neurosc Mtg Minneapolis Part I 253.

Takenaka T Inoue I Ishima Y and Horie H 1971 Excitability of the
    surface membrane of a protoplasmic drop produced from protoplasm

in <u>Nitella</u> <u>Proc</u> <u>Japan</u> <u>Acad</u> <u>Sci</u> <u>47</u> 554-557.

Vallery-Radot P 1922 <u>Oeuvres</u> <u>de</u> <u>Pasteur</u> Tome II Masson et Cie, Paris 328.

Wald G 1954 The origin of life <u>Sci</u> <u>American</u> <u>191</u>(2) 44-53.

Wigner E 1961 The probability of the existence of a self-replicating unit. In: E Shils, Ed <u>The</u> <u>Logic</u> <u>of</u> <u>Personal</u> <u>Knowledge</u> The Free Press Glencoe IL 231-238.

Woodby C D 1982 <u>Memory,</u> <u>Learning,</u> <u>and</u> <u>Higher</u> <u>Function</u> A Cellular View. Springer-Verlag, New York 121.

Yockey H P 1981 Self organization origin of life scenarios and information theory <u>J</u> <u>Theor</u> <u>Biol</u> <u>91</u> 13-31.

Yuki A and Fox S W 1969 Selective formation of particles by binding of pyrimidine polyribonucleotides or purine polyribonucleotides with lysine-rich or arginine-rich proteinoids <u>Biochem</u> <u>Biophys</u> <u>Res</u> <u>Commun</u> <u>36</u> 657-663.

MICROSCOPIC-MACROSCOPIC INTERFACE IN BIOLOGICAL INFORMATION
PROCESSING

Michael Conrad

Wayne State University

Detroit, Michigan   48202

SUMMARY

Recent experimental work suggests that chemical messengers
associated with the neuron membrane serve as a link between macroscopic and microscopic information processes in the brain. Arguments
based on the physical limits of computing, on computational parallelism, and on evolution theory, suggest that microphysical computing
processes enormously enhance the brain's computing power. A number
of models are briefly sketched which illustrate how molecular
switching processes could be recruited for useful biological
functions. The flow of information between microscopic and macroscopic forms is suggestive of processes which occur in a measuring
apparatus, and the implications of this analogy are considered.

I.   INTRODUCTION

It is commonly assumed that neurons play an important role in
biological information processing. There are a large number of neurons in the human brain--at least 10^{10} large ones and 10^{11} small
ones. Glial cells are even more plentiful, but neurons have attracted
a great deal of attention for at least two reasons. The nerve impulse
(with its on-off character) is a clear means of information transmission in peripheral neurons. The second reason is that in the

93

brain neurons are tied together into elaborate circuits. This has led to the natural idea that the neuron is a kind of switching element, analogous to the switching elements of a computer. This paradigm received its most significant impetus from McCullough and Pitts' demonstration in the 1950's that networks of formal neurons (threshold elements with on-off firing behavior) can implement any computing function whcih can be realized with a digital computer.[1]

There are, of course, a number of direct clinical reasons for believing that neurons play an important role in processes such as memory, learning, language, and thought. But it is important to recognize that this evidence does not establish the nature of this role. That is, it does not establish that biological information processes such as pattern recognition and problem solving are computations mediated by the on-off (or even the graded) firing properties of neurons.

In this paper I will argue that, for the most part, the nerve impulse plays a rather different role in biological computing. The argument is based on recent experimental discoveries about the effect of cyclic nucleotides on nerve activity. Cyclic nucleotides (cAMP and cGMP) are second messengers which link macroscopic signals impinging on the cell (mediators and hormones) to molecular events inside the cell. It is now known that these second messengers can also link the biochemical processes inside the neuron to electric potential changes of the membrane and to the production of spike potentials. The possibilities opened by this discovery are profound. The classical neurophysiologists tended to ignore the biochemical switching processes inside neurons because they found it difficult to understand how these switching processes could be effectively coupled to the electrical switching activity which either resulted from sensory input or controlled motor output. As a consequence the doctrine that this electrical activity provided the substrate of thought became popular, despite the fact that the nerve impulse is highly dissipative and slow in comparison to molecular switching processes.

I shall argue (in light of the new discoveries) that information processing in the brain is a vertical affair involving a flow of information from macroscopic to microscopic form, and back to macroscopic form. The nerve impulse is an intermediate, but relatively macroscopic stage in the hierarchy. A spoken word or a written symbol is a manifestly macroscopic stage. This is converted into a nerve impulse on the input side, preprocessed in a relatively macroscopic fashion at the level of neurons, linked via the cyclic nucleotide system to the biochemical and molecular layers of computing inside the neuron, converted back into a nerve impulse via the cyclic nucleotide system, and finally converted into manifestly macroscopic behavioral acts, such as sound production or other motor activities. Nerve impulses may still serve a computing function in this framework, but their primary function is to provide one stage of a communication link between microscopic computing processes in different neurons or as one stage in the communication link between microscopic computing processes and the outside world.

II. THE VERTICAL SCHEME OF BIOLOGICAL INFORMATION PROCESSING

The framework is illustrated in Figure 1. Computation occurring above the dashed line is mediated by the electrical activity of neurons. Information processing above this line is essentially a horizontal affair, involving macroscopic switching events in networks of neurons. The new feature is the chemical and molecular layer of activity below the dashed line. The nerve impulse triggers the production of chemical signals, which in turn control molecular switching processes. Protein enzymes, nucleic acids, membrane components, and other intracellular structures, such as microtubules and microfilaments, are the media of computation. After the molecular computation is completed, chemical messengers are produced. These messengers can either trigger new molecular computing events or trigger the firing of the neuron. Physically, the dashed line corresponds to the neuronal membrane, with nerve impulses

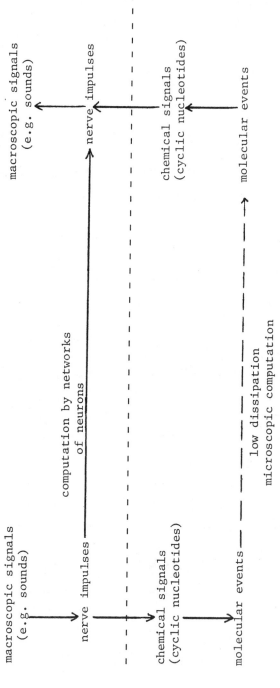

Fig. 1. Vertical scheme of biological information processing. The interfacing of macroscopic and microscopic events is mediated by cyclic nucleotide reactions of the type sketched in Fig. 2. The dashed arrow connecting the initial and final molecular events associated with microscopic computation is intended to indicate that microscopic computations could be linked to each other by chemical signals before being converted to the nerve impulse.

corresponding to macroscopic excitations of this membrane. Chemical
signals controlled by and controlling the nerve impulse are produced
on the membrane.

The important feature of the vertical scheme is the conversion
of information from a macroscopic representation to a microscopic
representation and back to a macroscopic representation. The most
macroscopic representations are the input from the environment and
the motor output of the organism. The external input is converted
to the relatively macroscopic nerve impulse, which is in turn converted to a chemical signal. The chemical signal is much less
macroscopic than the nerve impulse, but it still involves a statistical aggregate of molecules. The molecular events controlled by
this chemical sign are yet more microscopic. These molecular events
must occur either in protein enzymes or must be coupled to the action
of protein enzymes. This is necessary since chemical reactions must
be catalyzed in order to make the results of the microscopic computation macroscopic. These relatively microscopic chemical signals
trigger the relatively macroscopic nerve impulse. which in turn
controls the highly macroscopic motor actions of the organism.

Some information processing can take place at the macroscopic
level of neuronal switching, as in conventional brain models.
Certainly, sensory information is preprocessed by receptors and
processed as it is transmitted to central neurons. Processing of
information can also accompany the conversion from a macroscopic
representation to a microscopic representation. Different levels
of representation, such as the level of chemical reactions and
signals, can provide a substrate for computation, just as the
electrical and molecular layers can. But clearly the greatest
potential reservoir of computing power occurs at the molecular level.

The vertical scheme has a second important feature. Present
day digital computers are essentially built from switches which can
be either in one of two states, say 0 or 1. These switches are
simple. From the information processing point of view, their only
relevant properties are the two states. The switching elements in

the vertical scheme are much more complicated. An enzyme is a switch
in the sense that a chemical reaction is on, in its presence (or when it
is activated) and for all practical purposes off in its absence (or
when it is not activated). But in addition it is a pattern recognizer. It recognizes its substrate and recognizes specific bonds in
its substrate. It also recognizes control molecules which modulate
its activity. This pattern recognition is usually attributed to the
complementary fit between enzyme surface and substrate surface. The
short range van der Waal's interactions allow for a strong interaction only when the fit is precise. Thus, the enzyme is a powerful
tactile pattern recognizer.

This type of tactile pattern recognition is computationally
complex. That is, to simulate it with a conventional computer
(basically with a system of simple on-off switches) requires many
steps. In fact, this is just the kind of problem which computers
are so far not very successful with. The enzyme is, in effect, an
intelligent switch.

In reality, of course, the physical processes which accompany
substrate recognition and catalytic action are much more subtle than
captured in the idea of complementary fit. Chemically, the action
of the enzyme is accompanied by a series of electronic and nuclear
rearrangements. It is likely that some of these rearrangements are
associated with quantum mechanical instabilities which allow for the
transient opening of highly selective potential barriers.[2] The
important point (for the present purposes) is that the enzyme performs a complex information processing task, where complexity is
measured by the number of steps which a conventional computer would
require to perform a comparable task. In this sense, computation in
the microscopic representation is mediated by extremely powerful
computing primitives.

III. CYCLIC NUCLEOTIDE SYSTEM AND EXPERIMENTAL SITUATION

The idea that computing takes place below the dashed line (at
the chemical and molecular level) has been resisted by many

neuroscientists on the grounds of awkwardness. How could the manifestly present electrical layer of switching activity be coupled to
molecular switching processes? In recent years, the situation has
been reversed. Chemical coupling and signal mechanisms have been
discovered in a variety of CNS neurons. Indeed, it might now be
said that the ubiquity of these mechanisms makes it awkward to omit
the molecular layer from our model of CNS information processing.[3]

   To understand how these coupling mechanisms work, it is useful
to review some basic features of cyclic nucleotide biochemistry.
The ideas were originally developed by Sutherland in the context
of hormonal systems.[4] Peptide hormones secreted into the body fluids
do not directly enter into the cells they control. These hormones act
on receptors on the cell membrane to produce second messengers.
Cyclic AMP (cAMP) is one such second messenger. The hormone (or
first messenger) triggers the production of cAMP (the second messenger) which in turn acts on a trigger protein inside the cell. The
trigger protein is called kinase. Actually, there is a family of
kinases. The kinase triggers an effector protein by stimulating
phosphorylation. Since there are a variety of kinases, the second
messenger can have different effects in different cells. Separating
the kinase from the effector protein confers a great deal of evolutionary flexibility, which is consistent with the fact that the same
basic cyclic nucleotide control system is used for radically
different forms of control in different cells.

   In reality the cyclic nucleotide system is much more complex
than this. The interaction of the first messenger with the receptor
on the membrane stimulates the production of cAMP by activating the
enzyme adenylate cyclase. Another molecule, cyclic GMP (cGMP) also
serves as a second messenger, but the enzyme which produces it
(guanylate cyclase) is located throughout the cell. Binding proteins
are present which may control the diffusion rates of cAMP and cGMP.
Calcium plays an important role in the system. In the presence of
calcium, the enzyme phosphodiesterase converts cAMP and cGMP into

inactive mononucleotides. Still another enzyme, phosphoprotein
phosphate, reverses the action of the kinase on the effector protein.

The ubiquity of the cyclic nucleotide system in cellular control
processes has led many investigators to study its possible role in
central nervous system neurons. Possible roles (proposed by
Greengard[5]) include regulation of neurotransmitter synthesis, regulation of microtubules, and regulation of membrane permeability. The
sequence of events supposed to be involved in the control of membrane
permeability is illustrated schematically in Figure 2. The neurotransmitter produced by the presynaptic neuron serves as the first
messenger. The effector protein is a gating protein which controls
membrane permeability, that is, controls the nerve impulse. If the
permeability change is sufficient, the spike potential occurs. This
triggers an influx of calcium which deactivates the cAMP and, therefore, resets the neuron.

There is a considerable amount of experimental evidence which
supports the validity of this model. Some of it is based on the
external application of cAMP to various neurons.[6] But reasonably
direct experimental evidence has also been obtained by microinjection
of cAMP and cGMP into snail central neurons. This type of work has
been carried out by a number of groups. I would like to mention
briefly the series of experiments performed by the Liberman group
in Moscow, in part because they did some of the earliest work and
in part because I has the pleasure of working with this group during
an interacademy exchange visit to Moscow.[7]

The main result is that cAMP injected into large neurons of the
snail Helix lucorum usually depolarizes the neuron. That is, the
neuron fires more rapidly. A silent neuron may be induced to fire.
The effect is prolonged if a phosphodiesterase inhibitor is added,
one indication that the effect is not an artifact. Microinjection
of cGMP usually results in hyperpolarization, sometimes preceded by
a fast transient depolarization. Simultaneous injection of cAMP and
cGMP reduces the effect on the neuron, another indication that the

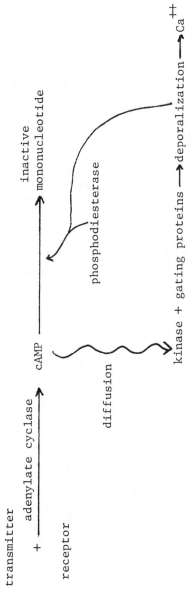

Figure 2. Cyclic nucleotide reaction scheme. Only the reactions involving cAMP are shown. The enzyme phosphoprotein phosphate (also not shown) reverses the action of kinase.

result is not an artifact. In some cases, injection at different
points on the membrane produces a different neuronal response,
possibly because diffusion of the cAMP from point of injection to
point of action is important. The time betwen injection and neuronal respose can sometimes be up to 1 to 3 seconds in large neurons,
again indicating the importance of diffusional or hydrodynamic
transport of cAMP. The great majority of neurons in the brain are
small (less than 50 microns in diameter), but techniques have not
yet been developed for these microneurons.

   Other investigators using the microinjection technique have
also found that cAMP and cGMP affect the electrical activity of
neurons. A number of different effects have been observed, with some
authors reporting that cAMP produces a long lasting hyperpolarization.[8] The reasons for the differing result is not entirely clear
at the moment. However, it is clear that cyclic nucleotides do
affect neuron electric activity. They can, of course, have a variety
of different effects, depending on what kinases and what target proteins are present. Indeed, the whole system can be used in entirely
different ways in different neurons. For example, the cyclic
nucleotide system is believed to be operative in receptor cells in the
retina. But instead of activating cAMP production, the photon
appears to activate phosphodiesterase production, leading to a local
reduction in cAMP concentration. For the present purposes, the
important experimental fact is that the cyclic nucleotide system
provides a channel of communication between electrical and molecular
layers of activity.

IV. PHYSICAL AND COMPUTATIONAL CONSIDERATIONS
   At this point I want to turn to three theoretical arguments.
The first involves the physical limits of computation. In principle,
these limits allow much more computation than can be achieved by
macroscopic computers. Indeed, the laws of physics are now known
to provide a "reservoir of computing power" which is completely
untapped by present day machines and by purely electrochemical models

of the nervous system.   The second argument comes from the theory of
computational complexity.   It involves the potential advantages of
parallelism.   The third argument comes from evolution theory.   It
is concerned with the type of organizations which allow for effective
utilization of the computing reservoir and of parallelism.   All these
arguments developed independently of cyclic nucleotide biochemistry.
They revealed in principle that the physical limits allow for an
immense amount of computation in comparison to those invoked in the
classical models.   But this immensity has for the most part been
viewed as a theoretical curiosity, a kind of physical esoterica.
The reason, I believe, is the lack of any plausible mechanism for
harnessing it.   But the elucidation of cyclic nucleotide communication systems in neurons alters this situation radically.

A.   Physical Limits of Computation

In recent years an important theoretical result has been obtained on the physical limits of computing.   This is the Bennett-
Fredkin result that there is no hardware-independent limit on the
energy dissipation required for computation.[9]   To appreciate this
result, it is useful to consider how physical limits bear on molecular computing and then how the Bennet-Fredkin result bears on these
limits.

First some numbers.   Neurons and transistors dissipate about
10^{10} kT/step.   Enzymatic reactions involving DNA and RNA involve
about 10 to 100 kT/step.   The switching time of computer switches
can be in the nannosecond range, as compared to the 10^{-4} to 10^{-3}
second reaction time of enzymes.   However, it must be remembered that
the complex pattern recognition task performed by the enzyme may
correspond to may computational steps on the part of a present day
digital computer.   The constituent electronic and conformational
events which contribute to overall recognition, control, and catalytic function of an enzyme are extremely fast, low dissipation
processes.

Now consider the physical limits per se. The main ones are:
i. The speed of light. The finiteness of light velocity imposes
a size limitation on computers. Increasing switching speed is of
no use if the speed of the computer is limited by the delay of signal transmission between switches. This is a practical limitation
in some current supercomputers, such as the Cray computer. This
computer has a cylindircal architecture in order to minimize the
distance over which signals must travel. Obviously, consolidating
computational work into structures of molecular dimensions would be
advantageous from the standpoint of this limitation.
ii. Fluctuation phenomena. As the dimensions of a computing primitive (e.g. enzyme or two state device) becomes smaller, it might be
expected to become more sensitive to errors arising from thermal
noise or from damage due to random bombardment by cosmic rays. This
is, in fact, a problem with some modern day chips. However, this
limitation may not be as severe as it appears for enzymes or other
macromolecular structures. This is due to the fact that the
switching elements of present day computers are macroscopic devices
in which the transition from state to state is fundamentally a phase
transition in that it involves the collective behavior of a large
number of particles. Some authors (notably Pattee[10]) have pointed
out that the discrete quantum levels which are apparent in single
macromolecules may enhance the reliability of molecular switches.
iii. Quantum limits. About twenty years ago Bremermann[11] showed
that the uncertainty principle together with the speed of light
imposes fundamental limits on the amount of computation which can
be performed by any system which operates in the fashion of a digital
computer. The argument is basically that the number of distinquishable states which could be available for information storage and
transmission is limited by the mass of a system. The amount of
digital computation potentially possible therefore has an upper limit
which depends on the mass of the universe, the speed of light, and
Planck's constant. For reasons which will become immediately

apparent, the Bremermann limit is many, many orders of magnitude better than that achievable with present day computing components.

iv. Thermodynamic limits. The major limit on present day computing systems is connected with energy dissipation, that is, with the heat of computation. Each computational step is accompanied by the conversion of electrical energy to heat energy. As the computer becomes more powerful, it becomes increasingly important to incorporate water or air cooling systems which carry this heat away. The rate at which a computing system can be cooled is expected to be the main limiting factor for the next generation of computers. (The main attraction of Josephson juction technology is that it should reduce this problem.)

A very natural question is whether there is a minimum heat of computation. This issue was considered by John von Neumann, a pioneer of computer science as well as other fields.[12] Von Neumann argued that each step of a computation was similar to a quantum mechanical measurement and must, therefore, require at least 0.7 kT of energy dissipation. This leads to a physical limit on computing which at normal temperatures is many orders of magnitude less favorable than that estimated by Bremerman. This limit is consistent with experience since 0.7 kT is ten orders of magnitude lower than the energy dissipation accompanying each computational step in a present day computer. The surprising discovery of Fredkin and Bennett is that von Neumann's limit is incorrect. It is possible in theory to construct switching elements with arbitrarily low dissipation and to design general purpose computers with these switching elements which also have arbitrarily low dissipation.

The intuitive consideration is that one can picture each switch in a digital computer as potentially jumping from one potential well to another at any given computational time step. There is no reason to treat the transition as a measurement. Dissipation may contribute to the speed and reliability of a computation. But there is no minimal amount of dissipation necessary in order for a computation to to occur independently of issues of speed and reliability.

Taken out of context, this conclusion appears rather trivial.
The problem is that if a computation can occur with arbitrarily low
dissipation (however slow and unreliable) this computation would be
thermodynamically reversible. But, in general, computation is
logically irreversible. For example, suppose that a computer
performs an inclusive "or" operation. Given the resulting value of
T, it is impossible to know if the previous values were (T,T), (T,F),
or (F,T). The nontrivial point is that this logical irreversibility
does not imply thermodynamic irreversibility. The reason is simply
that it is always possible to store the information about the past
states and then to use this stored information to run the entire
switching process backwards. Bennett even showed that it is possible
to construct an idealized general purpose computer which uses memory
in this way to approach thermodynamic reversibility. But, of course,
such a computer drifts randomly through its computation and its
memory requirements beome enormous. As soon as it becomes necessary
to use the memory store for some other computation, it must be
erased, which is definitely dissipative.

We can throw some new light on the Bennett-Fredkin construction
by looking at it from the point of view of the theory of computation.
According to the Turing-Church thesis, any function which is effectively computable is computable by a digital computing process
(ignoring time and space bounds). If a process could occur in nature
which is not computable by a potentially infinite digital computer,
this process could be used as a new computing primitive, thereby
invalidating the Turing-Church thesis by enlarging the class of
effectively computable functions beyond those computable by a digital
process. It is not possible to prove the Turing-Church thesis, since
the concept of effectiveness is an intuitive notion. The acceptance
of this thesis by computer scientists rests on the fact that all
alternative languages for describing algorithms which have ever been
proposed (Turing machines, Post tag processes, recursive functions,

Algol, Fortran, Pascal, PL1,...) are equivalent as regards the class
of functions which they are capable of computing. But acceptance
of the Turing-Church thesis implies a certain kind of equivalence
between continuous dynamics and the string processing operations
which define computation. It is always possible to simulate (however
inefficiently) any dynamical process with simple operations on
strings, that is with operations which are associated with computation.

So far, as is presently known, the basic dynamical processes
of physics (aside from measurement) are time reversible. For
example, the Dirac equation is time reversible. Suppose that it were
in principle impossible (even in the limit of idealization) to construct a thermodynamically reversible realization of a string
process. Then the dynamical process would have a property not
capturable by any string (or digital) process. But such a property
could then be used as a computing primitive which could enlarge the
class of effectively computable functions, in violation of the
Turing-Church thesis. Thus, thermodynamically reversible computation
must either exist in principle, or the Turing-Church thesis must
fail. If the Turing-Church thesis is admitted, thermodynamically
reversible computation must be as well. (Alternatively, demonstrating the impossibility of thermodynamically reversible computation
would provide a counterexample to the fundamental tenet of computa-
bility theory.)

The main significance of the in-principle possibility of the
thermodynamically reversible computation is that the amount of
computation which is theoretically possible is much greater than that
achieved by macroscopic computers or by neural networks in which the
computational events are confined to the electrical layer of activity.
I want to emphasize that this conclusion does not depend on the truth
or falsity of the Turing-Church thesis. If a counterexample to the
Turing-Church thesis were ever found (I do not exclude this possibility), the computational potentiality of nature would then be even

greater than that implied by our reduction argument (or by the
Bennett-Fredkin construction).

B. Computational Complexity and Parallelism

The computational complexity of an algorithm is usually defined
as the number of computational resources required for its solution.[13]
A computational resource may be a time step or a processor. A
processor is one of the physically realizable primitives, such as a
switch. For the present purposes, it is useful to think of algorithms as divided into two broad classes. The first are the
polynomial function of problem size. For example, grade school
multiplication requires on the order of n^2 steps, where n is a
measure of problem size (which is nonarbitrary up to order of magnitude). In general an algorithm is classified as polynomial time if
its growth rate is described by a function of the form $f(n) \approx n^m$,
where $f(n)$ is the maximum number of steps required. If m is small
(say m = 2 or 3), the algorithm may be called efficient in the sense
that the number of computational resources required to solve it does
not grow explosively as problem size increases.

The exponential time algorithms are defined by the fact that
$f(n)$ is not bounded by a polynomial (for example, $f(n) \approx 2^n$). Problems such as the travelling salesman problem and the quadratic
optimization problem probably belong to this class (the complexity
of a problem is usually defined as the growth rate of the fastest
algorithm for solving it). Exponential time algorithms are inefficient in the sense that the number of computational resources which
they require blows up exponentially with problem size. The combina-
torial explosiveness of these problems makes them intractible.

The increase in problem size which can be handled as switching
speed increases is easily ascertained if $f(n)$ is known. If
$f(n) \approx n^2$ a speedup of 10^{10} allows problem size to increase by at most
a factor of 10^5 (and still be solved in the same time). If $f(n) \approx 2^n$
the same increase only allows a maximum additive increase of
$\log_2 10^{10} = 33$. It is clear that even enormous speedups have only a
very small impact on the size of the problems that can be handled if

the algorithm is inefficient. But if the algorithm used by the
problem is efficient, speedup is enormously important. The important point is that increasing the computing power available to a
system is enormously significant if the problems which it must solve
have polynomial time algorithms (or are approximately or usually
solvable in polynomial time).

Now we are in a position to consider the significance of increasing parallelism, that is, increasing the number of components
active at any given time. To make matters concrete we can picture
a computer built from a collection of threshold elements, or unin-
telligent switches. A computer program can be hardwired into such
a collection of switches according to a definite construction rule.[14]
A number of different types of parallelism are possible in such a
system. One is component parallelism. If the program is not
inherently sequential, it is possible to trade time of computation
for number of components. One can reduce such computations to a
small number of time steps, but at the cost of introducing a very
large number of components. The potential parallelism, thus, becomes
very large, but the actual utilization of components in parallel is
not high. This inefficiency is the cost of programmability. It is
possible to prescribe the rule which a computer will follow only if
the possibility of conflicts among the components is completely
eliminated. Thus, the computational work is distributed among the
components in an extremely uneven fashion. But the minimum advantage
of parallelism is clearly obtained when the computational work is
distributed among the components as evenly as possible.

A second type of parallelism might be called logical parallelism.
For example, to even out the use of components one may contemplate
running different parts of a program on communicating machines. If
the possibility of conflict is excluded, one, in general, finds that
most machines are idle most of the time. If one allows conflicts,
the potential utilization of resources increases, but at the expense
of loss in programmability. One can still program such a system by
setting its initial state, but it is very hard to know or even write

down what program has been imparted to it. In general, the
theorems in this area make rather weak assertions. One can only
say that if one is lucky it is possible to obtain some maximum
possible speedup. But clearly, if one could obtain even some
fraction of an n-fold increase in speedup with an n-fold increase
in parallelism the increase in computing power would be highly
significant.

Actually, it is possible to do better than this, though not
with any system engineered like a present day computer. Most of
the interactions in a present day computer are suppressed. Each
component has an isolated dynamics which is parameterized on
inputs from a limited number of other components. The addition
of a new component does not affect these dynamics. But, in
general, there are n^2 interactions in a system of n particles
(according to currently understood force laws). In order to
obtain a programmable computer the engineer suppresses most of
these. Allowing more than one component to be active at any one
time in such a computer makes it difficult for the programmer to
know what program he actually imparts to the system when he sets
the states of the components, but it does not alter the isolated
character of these dynamics. This is why speedups of order n^2
never occur in conventional analyses of parallelism. But if all
the n^2 interactions in a system of n particles are turned on,
the difficulty of simulating the system goes up as n^2. If the
system performed an information or control function and each of
the n^2 interactions were relevant to the effectiveness of this
function, the potential increase in computational power would
be n^2. Of course it is unlikely that all the interactions could
be so effectively recruited. But even a modestly effective
recruitment efficiency would have an enormous impact on the
complexity of the tasks that could be performed.

The importance of effective recruitment is to some extent
illustrated by an electronic analog computer. The problem is

to compute a differential equation. All of the electrons in
the analog computer are effectively (evenly) utilized for the
solution of this differential equation, whereas most of the
particles in a digital computer programmed to solve the same
equation are rather inefficiently (unevenly) utilized. The
misleading feature of this example is the existence of simple
analogies which can serve as a guide to effective recruitment.
In general it is extraordinarily difficult to guess how a
parallel system should (or does) use its resources.

If a small fraction of the 10^{11} neurons in the human brain
could be effectively utilized for computing, the increase in
problem size which could be handled would be dramatic (for those
problems which admit an efficient algorithm). Inside each neuron
there are a comparable number of enzymes and other intelligent
switches. If one were lucky enough to recruit a small fraction
of these the impact on problem size would be enormous.

C. Evoluability versus Programmability

I am now going to argue that biological systems are built to be
lucky. More precisely, they are built for learning and evolution,
but at the cost of giving up programmability. First consider why
present day digital computers are not built to be lucky. The powerful feature of computer languages (Algol, Pascal, Fortran,...) is
that when we think of an algorithm, or effective process, we can be
sure that we can use the language as a means of expressing it. Any
effective algorithm that we can think up can be expressed in terms
of the primitives of any general purpose programming language, the
only difference between such languages being that their primitives
may be more or less suited to expressing the algorithm. Now suppose
that we alter the program by making a single change at random. The
common and often painful experience is that the program does not do
anything useful. If one waits for some superlanguage to be invented
which removes this delicacy one will wait for a very long time. Such
a language is theoretically impossible. If it were in general possi-

ble to predict in advance whether an altered program will even give
a defined computation, this would be tantamount to solving the famous
halting problem. But the halting (or self-computability) problems is
unsolvable. The argument is that for the halting problem to be solvable a program must exist which halts whenever a second program, taken
as data, does not halt. But the existence of such a program is contra
dictory when applied to itself since it would 'then halt only if it
does not.[15]

The situation is basically the same if a single random change
is made at the hardware level. The program is communicated to the
computer by using a compiler to set the states of the switches. The
situation is not altered at all since the compiler only re-expresses
the program in terms of the hardware primitives actually present.
Alternatively, one can hardwire the algorithm by using a "hardwiring
compiler" to set the connectivity of the switches. In either case
the original program is directly coded by the connectivity structure
and the state of the switches. Such systems may be built to be fault
tolerant in the sense that redundancy is used to preserve function
in the face of small changes. But, in general, one cannot expect
new useful behaviors to result from either small or large changes
unless these have their source in the intelligence of the programmer.

The organization of biological systems is fundamentally different. Protein enzymes provide a paradigmatic example. The protein
enzyme is more than an intelligent switch. It is a gradually
modifiable intelligent switch. This gradualness property is connected to the fact that macromolecules such as proteins have two types
of informationally significant description. The first is the
sequence of monomers, such as the sequence of amino acids in proteins. The second is the three dimensional structure and charge
distribution which determines the specificity and catalytic activity.
This three dimensional structure arises from the linear arrangement
of macromolecules through an energy dependent folding process. If
the sequence of monomers is slightly altered, the three dimensional

structure and function determined by it is often slightly altered
as well. This is connected with the fact that the folding process
is a continuous dynamic process. The introduction of continuity
introduces the possibility of small perturbations. In contrast, all
continuous dynamics are suppressed in computing systems. All levels,
from symbolic program to machine structure, are completely discrete.
The notion of a small perturbation in such a system is not in general
definable. This is an implication of the halting problem argument.
It is impossible to put any reasonable metric on the change in behavior exhibited by a computer program or digital computing device in
response to small changes in its structure.

Many other examples could be presented. The neuron itself is
an intelligent switch whose dynamics can be modified in a gradual
manner. The developmental system of the organism as a whole has this
gradualism property. The importance of this property is that it
allows for evolution through variation of structure and selection of
useful structures. Without the gradualism property the likelihood
of a useful structure arising from a single random change is extremely small. If the system operates like a computer, the set of
undefined programs which can result from such changes is enormously
larger than the set of defined ones. If a computer is built out of
intelligent, modifiable switches, it can evolve because the switches
can evolve in a gradual manner.

I have made two claims. The first is that structurally programmable systems (systems whose structure codes the program which
generates their behavior) cannot, in general, evolve by variation and
selection. The second is that this restriction on evolvability does
not necessarily apply to systems which are not structurally programmable. The behavior of such systems may be generated by a program
(or may be governed by a map which embeds a program similar to a
computer program). But the program is no more written in the structure of the system than Newton's laws are written in a falling apple.
This capacity to have function written into structure is the powerful

feature of digital computers which allows us to communicate algorithms
to them. But it is an enormously costly property in terms of both
evolvability and computational efficiency. This is due to the fact
that we are forced to work with a relatively small number of simple
switches which may not be well suited for the particular task at
hand.

Actually, these claims have been expressed too strongly.
Computer programs can be written which learn through evolution. But
the class of mutations which is allowed must be highly restricted,
for example, restricted to continuously adjustable parameters.[16]
The tack of our group has been to enlarge the class of problems which
can be addressed in this manner by introducing continuous features
and various forms of redundancy. Clearly, the amount of computation
must increase. If enough computational power were available to
simulate biological systems completely, we should be able to duplicate their evolutionary properties. Thus, to make our claim accurate,
it is necessary to add the element of computational work. Evolution
comparable to that exhibited by biological systems can be achieved
with structurally programmable computers only at the expense of an
enormous amount of computational work. Biological systems are built
for evolution because they cleverly use the laws of physics to do
this work. An enzyme is built to recruit its "resources" for folding
in an efficient manner; but a general purpose computer which attempts
to achieve gradualism by simulating enzyme folding must use its
resources in a most inefficient manner. It is in this sense that
biological systems are built to be lucky.

V.   BRAIN MODELS

We can summarize the situation thus. Biochemical and neurophy-
siological experiments reveal a messenger system capable of linking
macroscopic and microscopic information processes. Physical arguments
suggest that microscopic processes can, in principle, be harnessed
to enormously amplify the reservoir of computing power available to

systems in nature. Computational arguments suggest that the effectiveness of this reservoir can be further enormously amplified if its
components are recruited in parallel. The evolutionary argument
suggests that this reservoir of potential computing power can be
effectively tapped through evolutionary mechanisms. But this is not
possible if the system structure is programmable in the sense that our
present day electronic computers are.

How might these features be incorporated into models of the
brain? There are, undoubtedly, numerous mechanisms which could lead
to functionally useful behavior. The following four illustrate some
of the possibilities:

i. Neuronal pattern recognition and generation. The cyclic nucleotide system in neurons provides a means for both spatial and temporal
pattern processing by individual neurons. The presynaptic pattern of
inputs to the neuron gives rise to a pattern of cyclic nucleotide
production. Liberman [17] has pointed out that the diffusion of cyclic
nucleotides from loci of production to loci of action can be used for
information processing. The diffusion process converts the pattern
of cyclic nucleotide distribution in the cell. How this is interpreted by the cell depends on the location of kinases on the membrane
or in the interior of the cell.

The neuron can also serve as a temporal pattern generator. If
the endogenous production of cAMP is high enough, the neuron will
fire spontaneously. The influx of calcium will convert cAMP to the
inactive mononucleotide, stopping the firing as soon as the gating
proteins are dephosphorylated. The endogenously produced cAMP then
begins to build up again and the process repeats. How such a neuron
will respond to a temporal pattern of presynaptic input depends on
its stage in this cycle. [18]

ii. Neuron as analog of effector cell. The pattern processing
capabilities of neurons can be utilized for the all important task
of controlling and coordinating motor behavior. Suppose that the
organism is presented with a complex sensory situation. The first

problem is to decide what sort of motor act to undertake. For
example, should it move from position A to position B or from position A to position C? Once this decision is made, it is necessary
to implement it by providing the muscle cells of the body with two
items of information. The first is, how much should they contract?
The second is, at what time should they contract? The human has
six hundred and thirty nine muscles and numerous muscle cells associated with each muscle. It is probable that in many motions only
a few muscles are critically involved. But even so, it is clear that
adjusting the two required items of information is an extremely
difficult optimization problem. The situation is further complicated
by the fact that the exact adjustment must depend on the details of
the sensory input (deciding on a backhand stroke in a tennis match
and actually executing a particular backhand are two different levels
of motor control).

How might cyclic nucleotide mechanisms play a role here? An
important fact is that the brain maps the body in the sense that the
anatomists can localize portions of the brain to which information
from specific sensory organs is transmitted. The motor organs of
the body are similarly mapped. It is, therefore, plausible to assume
that individual neurons or small groups of neurons have responsibility for specific muscle cells. To coordinate muscular activity, it
is, therefore, sufficient to transform the sensory information transmitted to the responsible neurons to suitable output in terms of
timing and frequency of firing. The cyclic nucleotide system is
ideally suited for this. Timing can be controlled by altering the
distance of the kinases from the presynaptic input or altering the
number of binding proteins present, thereby altering the diffusion
rate of cAMP. The frequency of firing can be controlled by altering
the number of dephosphorylating proteins. After the neuron fires,
the cyclic nucleotide concentration pattern is reset to the ground
state. But if dephosphorylating proteins (reverse kinases) are in
a low concentration, the gating proteins will remain activated and

the neuron will continue to fire. Many other proteins in this system, such as the kinases, the number of gating proteins, and the
distribution of phosphodiesterase, can also be adjusted.

In the case of innate motor behavior, these adjustments are made
through the processes of phylogenetic evolution. The beauty of the
system is that it is built for effective evolution. This is due to
the fact that it operates through two levels of dynamics. The higher
level is the diffusion dynamics of the cyclic nucleotide system.
This is gradually modifiable. The lower level is the dynamics of
the enzymes which interpret the diffusion patterns. This is also
gradually modifiable. Furthermore, these dynamical features provide
an intrinsic generalization ability. This is due to the fact that
similar input patterns will often give similar neuronal responses,
enabling the organism to respond appropriately to situations to which
it had not previously been exposed.

In reality muscle cells are subject to both excitatory and inhibitory controls, originating in different regions of the brain. Some
of the relevant input to the neurons is proprioceptive, that is,
comes from the muscles themselves. But this just means that the
analog entity consists of a number of different neurons taken as a
group rather than a single neuron.

iii. Selection circuits model. Some motor behaviors are learned
in the course of an individual's development. An ontogenetic adaptation algorithm must be used to adjust the various proteins in this
case. It is reasonable to consider an alogrithm that works on evolutionary principles, since the system already possesses features
which make it suitable for phylogenetic evolution.

One clue is that the brain is highly redundant. In some cases
up to eighty percent of the cortex can be removed without significant impairment of specific function. But the rate of learning
decreases.[19] We can think of these redundant regions as competing
populations. We suppose that different local networks control the
organism's behavior at different times. The differential performance

of these redundant networks is ascertained and the best ones are
used to trigger the production of the same proteins in the less fit
networks. Two mechanisms are, in principle, possible. Inducers from
the fit networks could be transported to the less fit networks; alternatively nucleic acid segments from the fit networks could be transported to the less fit networks. This model is called the selection
circuits model because it presupposes a higher level of circuitry
which controls which local networks are active and which records the
pleasure or pain connected with the performance of that network. The
model is hypothetical, but it has been possible to show (in collaboration with R. Kampfner) that learning algorithms based on this idea
are highly effective.[21] Interestingly, they become increasingly
effective as an increasing share of the responsibility for information processing is allocated to single neurons.

iv. Neural networks and reference neuron scheme. How does the view
of a single neuron as a powerful computing device fit in with the
elaborate and highly specific neuronal circuitry which exists in the
brain? The situation can, to some extent, be analogized to telephone
circuitry. There is an increasing tendency to distribute computing
over a large number of small entities which are computers in their
own right. The elaborate telephone circuitry connecting these
entities is all important for communication, but not too important
for computation. The circuitry of the brain can be interpreted
similarly, but with the all important difference that the molecular
switching primitives in the neurons must be selected for very specic-
fic information processing functions. This is necessary if the
tradeoff between programmability and efficiency is to be honored.
It is possible to honor this tradeoff without losing ontogenetic
flexibility if there is a large library of different types of neurons, each embodying a relatively fixed "program" in its internal
molecular and chemical dynamics. The programming problem for this
system is transformed to the problem of establishing suitable communication linkages between the relatively complicated programs
"canned" in different neurons.

It is not necessary to postulate the existence of any de novo
mechanism for establishing these linkages. The association mechanism
is sufficient. When an organism is presented with some sensory
situation, a series of neuronal events is triggered. At some level
in the brain, a collection of neurons or, at very least, a single
neuron fires. The problem of memory is to trigger the firing of a
similar collection. The problem of associating A and B (for example,
name and face) is to have A trigger the neural events coordinated
to B, or conversely. Communication channels are presumably established between the neurons whose firing is coordinated to A and those
which are coordinated to B. The same mechanism can be used to establish communication channels between different neurons responsible for
different information processing tasks. If these linkages are useful,
they can be fixed, just as some memories are transferred to long term
memory.

To make matters definite, I will sketch one possible memory
mechanism which would be adequate for this purpose. It falls into
the class of Hebbian, or cell assembly, models.[22] The distinctive
feature is that there are two levels of neurons, reference neurons
and primary. neurons. For the present purposes, the primaries may
be identified with the library neurons. Reference neurons are control neurons which contact a large number of primaries. When a
primary becomes activated, its synapses become modifiable. Any
reference neuron that fires at the same time will modify the synapses of all the active primaries it contacts. In the future, activation of any of these reference neurons will reactivate the pattern
of primaries active at the earlier point of time. A number of
additional features are required to make the reference neuron scheme
internally consistent and to allow for the development of different
types of memory structures. The important point is this. By rememo-
rizing altered patterns of primary activity under the control of
uncommitted reference neurons, it is possible to orchestrate the
library of different neuronal capabilities in different ways.

This picture is not alien to the experimental situation in neurophysiology. Features which are often emphasized are the existence of complex and hypercomplex neurons, the high specificity of neuronal circuitry, and the information carrying capacity of electrical and chemical outputs of neurons. The new feature is that the cyclic nucleotide system serves as a further communication link to a microscopic level of computing. The role of the horizontal communication links among neurons is to assemble the library of neuronal capabilities into functionally useful behavior. The downward role of the vertical communication links is to use the electric and chemical inputs to a neuron to prepare the state of the switches in the cell insofar as possible, thereby initiating the microscopic computation process. The upward role of the vertical links is to convert the end states of the molecular switching processes to macroscopic signals and thence to macroscopic behavior. The enormous potential parallelism of the microscopic processes must require a great deal of filtering in order to finally select out which of these processes controls macroscopic behavior, which must finally be reduced to one set of actions in the functional individual. It is probable that much of the elaborate structure of the brain subserves this filtering function.[23]

VI.   THE BRAIN AS A MEASURING SYSTEM

Now I wish to return to von Neumann's idea that computation and measurement are linked. There are two possibilities. One is that von Neumann erred in associating each step of a computation with a measurement process. The second is that what constitutes a computation is unclear, or at least becomes increasingly unclear as the system becomes increasingly microscopic. In order register the computation it is necessary to make a measurement. Thus, even a "reversible" computer must undergo at least 0.7 kT of energy dissipation at the end of its computation, since its wavefunction must be reduced in order to register the measurement. We can suppose that

the computer remains in a superposition of states up to this point
of registration, but viewing it as operating on a discrete time scale
up to this point then becomes somewhat artificial. If one insists
that a time step must be bounded by measurements, von Neumann's
analysis remains intact after all, but with the provision that a
great deal of computational work can be done between measurements.

I have returned to the issue of measurement at this point since
there is an analogy between the measurement process and the
microscopic-macroscopic interfacing which occurs in the brain. To
make a measurement on a microscopic system, it is necessary to couple
it to a device which assumes a macroscopic state correlated to the
microscopic one. Alternatively, a measuring device could also be used
to prepare the state of a microscopic system. These are just the
capabilities possessed by the cyclic nucleotide system, with the
macroscopic to microscopic direction of information flow corresponding
to state preparation and the microscopic to macroscopic direction
corresponding to state recording.

Many processes in biology involve similar amplification of
of molecular events and control of these events by macroscopic features. Gene expression is the classical example. The analogy of
such processes to measurement has been the subject of a great deal
of discussion.[24] The chief issue, whether reduction of the wavefunction occurs, is fraught with well known conceptual difficulties.
In this respect, the significance of the analogy (and even its reality) is arguable. But the importance and fascination of these conceptual issues should not obscure the evidence that microscopic-
macroscopic correlation processes occur in the brain which in essential respects are similar to those which occur in measuring instru-
ments.

To grasp the potential significance of these processes, imagine
that we wish to simulate the behavior of a microscopic computer with
a macroscopic omputer, such as a present day digital computer. We
have a great deal of time and storage capacity at our disposal,

though not an infinite amount. We accept as an article of faith the
Turing-Church thesis that whatever is effectively computable is computable by our macroscopic computer, but with qualification that
we recognize the practical limitations on time and storage. We have
no doubt, for example, that the behavior of a falling apple is
computable. If it were not, we could use the noncomputable feature
to violate the Turing-Church thesis. But we recognize a computer
simulation of a falling apple or of any other process in nature is
forgetful in that the computer can never be apple-like. We argue
that this does not violate the Turing-Church thesis because we believe
that being apple-like is not relevant to computing.

The situation is somewhat different when we use our macroscopic
computer to simulate another macroscopic computer. For example,
suppose we wish to use our silicon computer to simulate a silicon
computer of different design, or even a future carbon computer. But
if we write a program which incorporates a representation of an apple,
it should be completely immaterial whether we run this program on the
silicon computer or on the carbon computer. Similarly, if we claim
that human though processes are programs in action, we are obliged
to admit that it is immaterial whether these programs are embedded
in a silicon computer or in the brain. If these programs constitute
the whole of what is essential for cognitive processes, a simulating
computer can, in principle, be perfectly faithful in all essential
respects even if its structure and constitution differ radically
from that of the brain.

Now suppose that we are further emboldened to simulate a molecular computer in which microscopic physical processes are known to
be relevant. We admit that the brain may be such a computer and that
the microscopic processes at the subcellular level require quantum
mechanics for their proper description. That is, we admit that these
processes are not picturable in a classical manner since such pic-
turing, in effect, constitutes a measurement. The situation we face
has been precisely formulated by Professor Dirac.[25]

> ...In the case of atomic phenomena, no picture can
> be expected to exist in the usual sense of the word
> "picture," by which is meant a model functioning
> essentially on classical lines. One may, however,
> extend the meaning of the word "picture" to include
> any way of looking at the fundamental laws which
> makes their self-consistency obvious. With this ex-
> tension one may gradually acquire a picture of atomic
> phenomena by becoming familiar with the laws of the
> quantum theory.

Taken as a whole, this passage is not consistent with the working
hypothesis that the brain can be adequately modeled with a macroscopic computer. This is despite the fact that it is consistent with
the in-principle possibility that we can use our computer to solve
the quantum mechanical equations which describe the microscopic computer. The problem is that the simulating computer is a macroscopic
system which is classically picturable, whereas the simulated system
is a microsystem which is not classically picturable. Thus, one of
the properties our simulation must forget is nonpicturability.

In the case of simulating a macroscopic system, we could always
argue that the forgotten feature was an inessential one not plausibly
relevant to computation. Now the forgotten feature is perhaps the
most fundamental feature of nature. To assume that the programs of
a classically picturable computer constitute the whole of what is
essential for the behavior of a nonpicturable system is not obviously
self-consistent. (The logic here is different from that in section
4(A), where we argued that reversibility should not be a forgotten
feature. Reversibility is a feature of the equations describing a
system and, therefore, it should be possible to duplicate this feature with string processing. Nonpicturability is more subtle,
because we admit that a mathematical picture exists and for the
purpose of the present argument we admit that this mathematical
picture can be perfectly duplicated with string processing operations.

What has happened is this. The Turing-Church thesis survives
the argument since it allows for infinitely long and infinitely large

computations. But as soon as we admit time and space bounds, we
have to face the computational cost of computing the behavior of a
nonpicturable computer with a picturable one. This cost is a new
one, additional to those connected with irreversibility, lack of
parallelism, and programmability. We, of course, recognize that a
general purpose macrocomputer can never use its computational
resources as efficiently as special purpose macrocomputers for suitable chosen classes of problems. Now we have to add that no macrocomputer can use its resources as efficiently as a microsystem for
classes of problems for which the nonpicturability of the microsystem
is relevant. The manner and extent to which information processing
can be enhanced by utilization of microscopic dynamics is an open
question. It cannot be ascertained by a backdoor argument of the
type presented here. But the ubiquity of microscopic-macroscopic
coupling mechanisms in living systems suggest that the special
features of very small scale phenomena do contribute to their remarkable information processing capabilities.

VII. CONCLUSIONS

We can summarize by considering how features of present day
computers compare to those of the brain. The computer is essentially
built from simple switches, though there is now a tendency to aggregate these into more complex primitives. The brain operates on the
basis of a hierarchy of switching and communication processes, with
protein switches at the microscopic level of the hierarchy and neuronal switching processes occuring at a macroscopic level. The
dissipation accompanying enzymatic switching is negligible so far
as the enzyme itself is concerned (it is a catalyst). Switching is
relatively slow, but this is offset by the sophisticated pattern
recognition that accompanies the switching process. The chemical
reactions controlled by these switches are more dissipative and the
dissipation accompanying the nerve impulse is higher yet, corresponding roughly to that of present day computer switches.

Present day computer switches are fixed building blocks, capable of evolving into new forms only through human intervention. The protein and neuronal switches of the brain are built for effective evolution, an unavoidable feature since they are the products of evolution. The communication between switches in von Neumann type computers (basically all commercially available computers) is highly structured for sequential access and retrieval. The communication between protein switches in the brain is mediated by chemical reactions and by nerve impulses triggered by these chemical reactions. Memory is highly distributed over many neuronal switches which act in parallel. The pattern processing capabilities of the neuron depend on its internal molecular switching mechanisms and the elaborate connectivity of the brain may be viewed as communication lines between a variety of specialized and extremely powerful information processing devices.

The dynamics of the computer is completely discrete. All elements of continuity are irrelevant. The enzymatic and neuronal dynamics in the grain has both discrete and continuous feature, a fact relevant to evolution, learning, and generalization. The timing of processes in the brain is inadequately understood. This is an important point since the occurrence of formal computation processes (essentially processing strings of symbols) is impossible in the absence of a discrete timing mechanism.

The major feature of present day computers is programmability. The brain is programmable in the sense that we can interpret (read and follow) rules. But it is not programmable at the level of structure. The loss of such structural programmability allows for effective evolution. Amenability to evolution involves some cost since it requires various forms of redundancy. But it allows for enhanced efficiency since components can be fashioned specifically for the tasks at hand. It also increases the likelihood that parallelism will be effectively utilized. The standardization of building

blocks in a computer allows for parallelism only of order n, where
n is the number of components. In principle, however, parallelism
can increase as n^2. Obviously, the number of switching elements in
the brain is enormously larger than that in a present day computer.
If even a small portion of this potential parallelism is utilized,
the impact on information processing power would be enormous.

The major feature of the brain is that it appears to harness
microscopic dynamics. Quantum mechanics is relevant to the dynamics
of switches in a computer, which can be completely understood in
formal terms. The situation is different in the brain due to cyclic
nucleotide mechanisms which provide a microscopic-macroscopic interface at the nerve cell membrane. Due to this interfacing, it is
highly likely that processes which are essentially nonclassical are
relevant to the macroscopic capabilities of the organism. The novel
situation is that the behavior actually executed would be completely
picturable in classical terms, but not explicable in such terms.

We have used the metaphor of a program at various points in this
paper to refer to the map (taken in the mathematical sense) governing
the dynamics of the brain. I think it is fair to say that this
metaphor has assumed paradigmatic significance in brain science and
in cognitive science generally. The features we have discussed (high
parallelism, continuity and timing properties, tactile switching
mechanisms, structural nonprogrammability, and the controlling role
of microphysical events) make it very unlikely that the dynamics of
the brain is directly generated by programs of the type with which
we are familiar. According to the Turing-Church thesis, programs
should exist which, in effect, compute these dynamics, that is,
in effect, compute the maps responsible for intelligence. But it
is not likely that such programs could be computed or even embedded
in any computer of present day design. In this respect, the difference between the physiochemical constitution and structural plan of
the brain and that of the computer is decisive.

ACKNOWLEDGEMENTS

This research was supported by grant MCS-82-05423 from the National Science Foundation.

REFERENCES

1. W.S. McCullogh and W. Pitts, A logical calculus of the ideas immanent in nervous activity, Bull. Math. Biophys. 5 (1943), 115-133. For an extended discussion see M. Minsky, Computation: Finite and Infinite Machines (Prentice-Hall, Englewood Cliffs, N.J., 1967).

2. M. Conrad, Electron pairing as a source of cyclic instabilities in enzyme catalysis, Phys. Lett. 68A (1978), 127-130.

3. The problem of reconciling molecular mechanisms with canonical knowledge in neuroscience is used as one of the examples in G.S. Stent, Prematurity and uniqueness in scientific discovery, Sci. Am. 227 (172), 84-93.

4. G.A. Robison, R.W. Butcher and E.W. Sutherland, Cyclic AMP (Academic Press, New York-London, 1971).

5. P. Greengard, Cyclic Nucleotides, Phosphorylated Proteins and Neuronal Function (Raven Press, New York, 1978),

6. F.E. Bloom, G.K. Siggins, B.J. Hoffer and M. Segal, Cyclic Nucleotides in the central synaptic actions of catecholamines, Advanded Cyclic Nucleotide Res. 5, (1975), 603-618.

7. E.A. Liberman, S.V. Minina, and K.V. Golubtsov, The study of metabolic synapse. I. Effect of intracellular microinjection of 3'5'-AMP, Biofizika 20 (1975), 451-456. For a more recent paper see E.A. Liberman, S.V. Minina, N.E. Shklovsky-Kordy, and M. Conrad, Microinjection of cyclic nucleotides provides evidence for a diffusional mechanism of intraneuronal control, Biosystems 15 (1982), 127-132.

8. S.N. Triestman and I.B. Levitan, Alteration of electrical activity in molluscan neurons by cyclic nucleotides and peptide factors, Nature 261 (1976), 62-64.

9. C.H. Bennett, Logical reversibility of computation, IBM J.
    Res. Dev. 17 1973, 525-532. C.H. Bennett, Dissipation-error
    tradeoff in proofreading, Biosystems 11 (1979), 85-92. For a
    recent discussion see P. Benioff, Quantum mechanical Hamiltonian
    models of Turing machines, J. Stat. Phys. 29, no. 3 (1982),
    515-546.

10. H.H. Pattee, The physical basis of coding and reliability in
    biological evolution, in: Prolegomena to Theoretical Biology,
    ed. C.H. Waddington (University of Edinburgh Press, Edinburgh,
    1968), p. 67-93.

11. H.J. Bremermann, Optimization through evolution and
    recombination, in: Self-Organizing Systems, ed. M.C. Yovits,
    G.T. Jacobi and G.D. Goldstein (Spartan Books, Washington, D.C.,
    1962), p. 93-106. For an updated treatment see H.H. Bremermann,
    Minimum energy requirement of information transfer and computing
    Int. J. Theoret. Phys. 21 (1982), p. 203-217.

12. J. von Neumann, Theory of Self-Reproducing Automata, ed. A.W.
    Burks, (University of Illinois Press, Urbana, 1966), p. 66.

13. M. Garey and D. Johnson, Computers and Intractability (Freeman,
    Ney York, 1979). For a discussion geared to biological problems
    see M. Conrad and A. Rosenthal, Limits on the computing power
    of biological systems, Bull. Math. Biol. 43 (1981), 59-67.

14. M. Conrad, Adaptability (Plenum Press, New York, 1983), p. 233.

15. For a discussion of the halting problem and unsolvability see
    Martin Davis, Computability and Unsolvability (McGraw Hill,
    New York, 1958).

16. A classic example is provided by A.L. Samuel, Some studies
    of machine learning using the game of checkers, IBM J. Res.
    Dev. 3 (1959), 211-229.

17. E.A. Liberman, Analog-digital molecular cell computer,
    BioSystems 11 (1979), 111-124. The role attributed to diffusion
    in the above paper is different than that suggested here. For
    discussions closer to the present one see M. Conrad and E.A.

Liberman, Molecular computing as a link between biological and physical theory, J. theoret. Biol. 98 (1982), 239-252, and the paper on diffusional mechanisms of intraneuronal control referenced in 7.

18. These statements about pattern processing which emerge from the Liberman experiments have been verified by computer simulation. I acknowledge collaborative work with K. Kirby (not yet published).

19. K.S. Lashley, Brain Mechanisms and Intelligence (Chicago University Press, Chicago, 1929) (Reprinted by Dover, New York, 1963).

20. M. Conrad, Evolutionary learning circuits, J. Theoret. Biol. 46 (1974), 167-188. M. Conrad, Molecular information processing in the central nervous system, part I: selection circuits in the brain, in: Physics and Mathematics of the Nervous System, ed. M. Conrad, W. Guttinger, and M. Dal Cin (Springer, Heidelberg, 1974), p. 82-107.

21. R. Kampfner and M. Conrad, Computational modeling of evolutionary learning processes in the brain, to appear in Bull. Math. Biol. This computational study used formal models of enzymatically controlled neurons which were designed primarily to suit the requirements of a learning model. We now interpret the physiology of these neurons in terms of cyclic nucleotide mechanisms.

22. D.O. Hebb, The Organization of Behavior (Wiley, New York, 1947). For details of the reference neuron scheme see M. Conrad, Molecular information processing in the central nervous system, part II: molecular data structures, in: Physics and Mathematics of the Nervous System, ed. M. Conrad, W. Guttinger, and M. Dal Cin (Springer, Heidelberg, 1974), p. 108-127; M. Conrad, Molecular information structures in the brain, J. Neurosci. Res. 2 (1976), 233-254; M. Conrad, Principle of superposition-free memory, J. Theor. Biol. 7(1977) 213-219.

For an interfacing of the reference neuron scheme with the
selection circuits model see M. Conrad, Complementary molecular
models of learning and memory, BioSystems 8 (1976), 119–138.

23. I acknowledge stimulating remarks by B.D. Josephson on the
possibility that the brain performs a filtering function. The
problem of accessing suitable programs in the stochastic
information processing model of H.M. Hastings (this volume) is
also pertinent to this point.

24. See for example H.H. Pattee, Physical problems of decision-
making constraints, in: The Physical Principles of Neuronal
and Organismic Behavior, ed. M. Conrad (Gordon and Breach,
New York and London, 1973), p. 217–225; For a recent review
see M. Conrad, Adaptability (Plenum Press, New York, 1983),
Chapters 2 and 3.

25. P.A.M. Dirac, The Principles of Quantum Mechanics (Oxford
University Press, Oxford, England, 1958), p. 10.

STOCHASTIC INFORMATION PROCESSING IN BIOLOGICAL SYSTEMS

II - STATISTICS, DYNAMICS, AND PHASE TRANSITIONS

Harold M. Hastings

Hofstra University
Hampstead, New York   11550

Abstract. We discuss the use of probabilistic automata and
networks of probabilistic automata as a framework for modeling
biological information processing and explore several consequences.
In particular, we use the Ising model and recurrence in Brownian
motion to argue that the brain uses its two-dimensional surface for
primary data access, and consider biological attacks on polynomial-
time, and nondeterministic polynomial-time problems, My talk and
this paper form a continuation of joint work with Richard Pekelney
(Hastings and Pekelney, 1982).

I. INTRODUCTION

My talk discussed joint work on the use of networks of probabilistic automata as models for biological informantion processing
(Hastings and Pekelney, 1982), and my recent extensions to include
some aspects of the dynamics, statistics, and phase-behavior of these
networks. In this paper I shall briefly review the main predictions
and conclusions on Hastings and Pekelney (1982) and discuss the above
extensions.

Most mathematical models of the brain, ranging from Turing (1936)
machines and networks of McCulloch-Pitts (1943) neurons to more recent

models (cf. Arbib, 1965; Conrad, et al., 1974; von Neumann, 1966)
are essentially deterministic. However, this strategy of modeling
the brain as a formal deterministic computer is incompatible with
several characteristics of the brain: in particular with its gradual behavior and nonalgorithmic learning (Conrad, 1976 a, b). In
addition, biochemical information processing in the brain involves
reaction and diffusion of very small amounts of neurotransmitters,
sometimes as small as a few molecules (Elliot, 1969; Taylor, 1974;
Schmidt, 1978). This suggests using random walks to model the underlying processes rather than their "ensemble averages," diffusion
equations (cf. Rossler, 1974). We therefor were led to consider
(Hastings and Pekelney, 1982) modeling biochemical information processing at the neuronal level by probabilistic automata (Shannon
and Weaver, 1948, see also Ashby, 1958; von Neumann, 1966; Burks,
1970; Paz, 1971). We propose that the intrinsically stochastic behavior of these automata is essential to biochemical information
processing, and that it is not a class of errors to be avoided (as
in Poisson failures in early vacuum tube computers), or removed
through evolution toward greater efficiency (von Neumann, 1966; Burks,
1970).

Classical and Probabilistic Automata

An ordinary, classical automaton is an ordered five-tuple,
which consists of the following:

(i) A finite set called an input alphabet; an input is a non-
empty word constructed from the input alphabet,

(ii) A finite set called the output alphabet, consisting of all
possible outputs,

(iii) A finite set of internal states,

(iv) A next-output function f: (input alphabet) × (internal
states) ----→(output alphabet), and

(v) A next-state function g: (input alphabet) × (internal states) ----→(internal states).

Stochastic automata are defined similarly, except that the ranges of the next-state function g are probability distributions of the output alphabet and set of internal states, respectively.   Each time a stochastic automaton receives a (one-letter) input, one output and one new state are generated according to these respective probability distributions.

In (Hastings and Pekelney, 1982) we showed that the following predictions were consequences of our basic framework of systems of probabilistic automata.

Gradualism (Conrad, 1974).

The brain typically displays small changes in behavior in response to small changes in inputs ("programs" and "data"); in contrast with the discontinuous or even catastrophic response of computers to such changes.   Two features of large networks of probabilistics automata, each connected to many other such automata (each of the approximately 10^{11} neurons in the brain is connected to about 10^4 other neurons), yield such gradual behavior.   First, small changes in the probability distributions generated by the next-output or next-state function, corresponding to small modifications in the probability of any given output or next state, cause only small changes in behavior of the system.   Secondly, even a large change in a rarely accessed (as measured, for example, by the probability distributions) automaton also cause only small changes in behavior.   Note that gradualism can occassionally be violated in both cases.

Learning

Our networks exhibit both memory-based and modification-based (nonalgorithmic, and hence not available to deterministic computers) learning described by Conrad (1976 a, b).   The latter mode of learning is a consequence of both the nonprogrammability of the stochastic behavior of probabilistic automata, and Ising-type phase transitions discussed below, and in Section 2.

Memory Management

Our model makes several predictions about memory management.
We shall describe one key prediction below: the primary addressing
scheme is two-dimensional. This largely explains the folded and convoluted structure of the brain. In addition, stochastic access yields
rapid overviews followed by increasing detail, as the number of randomly accessed neurons grows in size. Finally, stochastic access
allows the location of imprecisely specified data, a feature not found
on typical computers without sophisticated programming.

Efficiency

The above differences in behavior between networks of probabilistic automata and networks of deterministic automata challenge para-
doxical estimates of low efficiency (on the order of 10%) of the brain.
We shall discuss aspects of our resolution of this paradox in Section
2 (dimensionality of the brain), Section 3 (polynomial speedups), and
Section 4 (stochastic attacks on NP problems), below.

We used similar mathematics to model the genetic system, with
one important difference: the probability distributions in the genetic system are more highly concentrated, with only rare mutations altering the instructions for reading the genetic data.

We note that probabilistic Turing (1936) machines might be better
models for genetics and evolution than probabilistic automata. Roughly, a probabilistic Turing machine consists of a probabilistic automaton together with an arbitrarily long (as needed) tape upon which
the automaton may read and write. The tape is realized biologically
by DNA, reading by reproduction and transcription, and writing by mutations.

Silent Genes and The Simultaneous Mutation Paradox

We provided the framework to formalize an idea of S. Ohno (1972)
that silent genes serve as storage areas for point mutations which
might be individually deleterious but beneficial in combination, thus
facilitating the incorporation of these multiple-point mutations.

We shall discuss this phenomenon further below, as part of a discussion of biological problem solving.

Punctuated Equilibria (Gould and Eldredge, 1977).

Our genetic model provides a mathematical framework whose dynamics exhibits general stasis (gradualism) with occasional bifurcations in evolution as proposed by Gould and Eldredge.

In this paper I shall focus on aspects of Hastings and Pekelney (1982) and recent developments which are likely to be of interest outside as well as inside biology: the use of the Ising model (Section 2), biological problem solving and polynomial speedups (Section 3), stochastic attacks on NP problems (Section 4), and the different roles of stastistical samples in biology and physics (Section 5).

I am pleased to acknowledge R. Pekelney, and Drs. M. Conrad, R. M. May, O. Rossler, and E. Yates.

II. MEMORY IS TWO-DIMENSIONAL, FOLLOWING HASTINGS AND PEKELNEY (1982)

In hastings and Pekelney (1982), we argued that the primary data structure for memory access is a two-dimensional array of neurons. We shall sketch that argument here, because its further consequences provide a strong justification for using networks of probabilistic automata as a model. Two factors enter into this analysis: first, random signals used in random (Brownian) searching must be recurrent, and second, local interactions must add strongly enough to cause "phase transitions," and thus conformational changes to provide non-algorithmic and long-term memory.

To study further the first factor, we assume that dissipation of signals is rapid enough so that in the absence of positive feedback, the geometry of primary data access in the brain is well-modeled as an unbounded subset of Euclidean space. We propose that this structure is accessed "stochastically" (the term "random access" is appealing, but it has a special meaning in computer science), by random walks traced out by sequences of neurons transmitting "read" instructions. In order that the appropriate, but initially unspecified, areas of memory be accessed by such a random process, so that feedback can develop, random walks must be recurrent. This requires that the dimension of the "stochastically" accessed neurons must be at most 2.

In addition, we considered the Ising model (see R. Shaw's talk (Shaw, 1983) for work of Shaw and Little on the Ising model in brain modeling) as the most appropriate model for the required phase transitions. Because the overall, global connectivity of the brain is relatively low (there are about 10^{15} synapses connecting about 10^{11} neurons, so that each neuron is connected with about 10^{-7} of the other neurons) , we consider only short-range interactions. In this case, the dimension must be at least two for these local interactions to interact strongly enough to cause phase transitions.

In combination with our previous discussion, this yields a dimension of two for primary "stochastic" access. Thus evolution toward increased memory capabilities should lead to maximizing surface area, in good agreement with the layered, and highly convoluted structure of the surface of the brain. In addition, this role of surface area is consistent with Lorber's observations (cf. Lewin, 1980) that many hydrocephalics, whose brains have normal surface area, but sharply reduced volume function with superior IQ's. We therefore also note that the information-processing capability needed for survival of an individual member of a species may be only a subset of those used by all members of that species collectively in evolution. This phenomenon is also seen in the use of silent genes to achieve a polynomial speedup in the rate of simultaneous point-mutations, as discussed below.

III. BIOLOGICAL PROBLEM SOLVING I - POLYNOMIAL SPEEDUPS, FOLLOWING
       HASTINGS AND PEKELNEY (1982)

In this section we shall consider the first of two aspects of biological problem solving. Following Hastings and Pekelney (1982) we observe that silent genes can be used to obtain a polynomial time speedup in solving the problem of achieving simultaneous point-mutations. In the next section we shall consider the use of parallel probabilistic automata in obtaining adequate approximate solutions to possibly exponentially large problems. We view the genetic system as , in part, a biological computer for finding evolutionarily stable (Maynard-Smith and Price, 1973) phenotypes. In man and higher mammals, much of the required computation is performed in the brain, where again stochastic techniques can yield similar speedups.

We shall consider the following aspects of the genetic system: highly repeated genes of which normally one specific copy is expressed and the others are "silent"; regulator genes which control which copies of these genes are to be expressed; and, finally, mutations which affect both types of genes.

Consider the evolution of a new phenotype whose genotype differs from the genotype of its evolutionary ancestor by several point-mutations, each individually deleterious as are most mutations. Without some special mechanism for hiding these separate mutations until all of the required mutations are realized, evolution would proceed very slowly because all of the required mutations would have to occur simultaneously in some one individual. However, this problem could be avoided if all of the required mutations occured first in silent copies of these genes, and a later mutation of a regulator gene caused these new copies to be read (cf. Ohno, 1972). We had considered a population of size N of the original species, a mutation rate of p (usually, on the order of 10^{-6}) per locus per year, and a mutation made up of n point-mutations. Then the use of stochastic behavior of regulator genes can reduce the "computing" time from order $1/Np^n$ to order $1/Np$ generations, a polynomial speedup. This overcomes the near impossibility (Conrad, 1972) of simultaneous point-mutations.

IV. BIOLOGICAL PROBLEM SOLVING II - STOCHASTIC ATTACKS ON
      NP PROBLEMS

We now consider biological solutions to more complicated (exponential time) problems. Many optimization problems, for example the

traveling salesman problem, fall in the class of NP-complete problems.
The traveling salesman problem is the problem of finding a path of
minimum total cost which visits each vertex in a graph whose edges
are labeled by their costs. Thus the traveling salesman problem is
a special case of evolutionary path-optimization problems considered
in O. Rössler's talk (Rössler, 1983). Even the decision problem,
"does there exist a path of total cost less than C (C given and
positive)", is NP-complete. This decision problem is analogous to
the evolutionary problem of finding a sufficiently good strategy to
fulfill the food and habitat requirements of a member of a species.

These problems require exponential time for solution by ordinary deterministic computers (and are thus usually considered intractable), but only polynomial time for solution by nondeterministic
computers. Roughly, only polynomial time is required to check whether any guess (this is where nondeterminism enters) at a solution is
actually a solution. It is important to note that any time required
to guess the correct solution is not included in the definition of
nondeterministic polynomial (NP) time. It seems reasonable that many
optimization problems which do not admit efficient reduction into similar smaller problems are at least this hard, and thus that many biological problems fall into this class.

We shall describe here a concept of biologically acceptable solutions, and show how networks of probabilistic automata can rapidly
find these solutions. For a solution to be acceptable, it must first
work in the absence of competition (consider, for example, the description of searching in O. Rössler's talk), and then be able to
resist invasion by competing solutions (as in Maynard-Smith and
Price's evolutionarily stable strategies).

The first criterion, survival in the absence of competition,
should not cause combinational explosions in computing requirements.
In terms of Rössler's talk, local or step optimization should suffice.
In a discussion of NP-complete problems, many, including the traveling
salesman problem, admit polynomial-time algorithms which yield

solutions within some factor of the theoretical optimum. Therefore,
the first criterion does not demand specialized computational
resources.

The second criterion, that a solution be evolutionarily optimal,
may require additional resources, which we propose consist of stochastic information processing. In Rössler's terms, evolutionarily
stable strategies require path-optimization. (Note that the traveling salesman problem is a difficult problem in path-optimization).
We consider optimization problems in class NP - these admit polynomial-time solutions by nondeterministic machines. A nondeterministic
machine may be viewed (and philosophically realized) as an exponentially large collection of deterministic machines working in parallel
(thus a simulation on a sequential, deterministic machine requires
exponential time). These nondeterministic machines make all possible
choices in a nondeterministic algorithm. Alternatively, one may e-
quip a determistic machine with an oracle (Garey and Johnson, 1979)
which makes guesses at the required choices. A problem is in class
NP if some specific sequence of guesses by the oracle lead to a
polynomial-time solution. However, many sequences of guesses may
lead to no solution. If the density of satisfactory guesses is
large enough, a family of parallel probabilistic automata would find
a satisfactory solution. In the case where these automata are realized by the process of mutation and evolution, satisfactory solutions
in at most polynomial time should be attainable. In addition, these
polynomial time solutions can be realized in linear time by the first
speedup process. In the case of information processing in the brain,
the requisite family of automata is realized by the large number of
neurons and synapses in an individual brain, and more generally in
the brains of all members of a social species. This computational
performance would be impossible without the large amount of redundancy in (at least most human) brains. Because solutions to these
problems can be communicated among members of a social species, some
individuals may function at high levels with much less redundancy.

Our first obervation above challenges the naive low estimated efficiency of the brain, and our second obervation shows why the hydrocephalics with high IQ's noted above does not challenge the need for
stochastic information processing.

V.   CONSEQUENCES FOR REALIZATION - THE SMALL SAMPLE PHENOMENON
     IN BIOLOGY

   Many authors, most recently Noda (1982), have considered the
rarity of functional DNA sequences among all possible DNA sequences
(cf. also Maynard-Smith's (1970) protein spaces). In fact, if we
consider evolution as an unstructured searching problem, it appears
to be intractable. How then does eveolution solve this problem.
We shall propose an analysis which appears reasonable, and suggests
that the evolution of some complex organisms is almost inevitable.
Their calculations have suggested a paradox in the evolution of
complex systems, whose genotypes involve relatively long DNA
sequences.

   We shall propose a resolution of this paradox based upon the
dynamics of stochastic information processing. Recall, from our
previous discussion that stochastic attacks on any NP problems in
evolution may yield good solutions, but will typically yield only
one of many possible solutions. This effect is realized in several
ways.

   First, the DNA in a complex organism contains many genes, each
coding for a particular protein, or more generally for a particular
function. As in Noda (1982), we also assume that the DNA contains
instructions for locating and reading these functional genes. Thus,
any permutation of the locations of these genes, together with an
appropriate change in instructions for locating these genes should
yield (an at least equivalent) functional organism. There are exponentially many permutations available, for example, a set of 50 locations admits 50!, or about 10^{64}, such rearrangements. In this sense
any functional DNA sequence represents only one of many equivalent

possible sequences. The inclusion of such a factor in Ohno's calculations would remove many apparent computational barriers to the
evolution of complex organisms.

Note that in a statistical approach to biology, few of the many
possible states (cf. also Maynard-Smith's protein spaces) will be
actually realized. This contrasts with the use of ensembles in
studying the quantum mechanics of, for example, gases at ordinary
temperatures and pressures.

Two other factors also reduce the apparent computational barriers
to the evolution of complex organisms. The first is the large number
of silent genes apparently used to resolve the "simultaneous-mutation
paradox," as discussed above. In addition, gene duplication in evolution (Ohno, 1970), allows evolution to proceed by replacing sections
of DNA by known functional sections. Again, very little of the apparent DNA-space need be searched to obtain evolutionarily stable solutions. These ideas are implicit in observations of Ho and Saunders
(1979) that the information content of DNA sequences on the codon
level far exceeds the information content of the phenotype on the
functional level.

Although these considerations do not increase the proportion of
functional DNA sequences among all possible sequences, they do increase the probability that complex functional sequences will evolve.

VI.   CONCLUDING REMARKS

We have proposed the use of networks of stochastic automata in
biology and described several consequences. Although the underlying
dynamics at the molecular level may well be deterministic, we feel
that information processing at higher levels is better modeled stochastically. In addition, this stochastic behavior is better modeled
by random walks themselves rather than their "ensemble" averages,
diffusion equations. Thus redundancy, symmetry, and sampling must
be seen differently in biology and statistical mechanics.

We are currently undertaking simulation studies of networks of probabilistic automata, with G. Hastings, Dr. S. Waner, and C. Zingarino. These models embody both dissipation and positive feedback (cf. R. Ulanowicz, 1980). If, as diffusion processes suggest, dissipation is well-modeled by exponential decay, then the positive feedback must proceed at a faster-than-exponential rate for structure to develop generically. This provides an information-theoretic basis for multiple messengers and multistage catalysis. It also appears interesting to pursue brain and genetic involving towers of probabilistic (or ordinary) automata, each of whose next-state and next-output functions is controlled by the output of a higher level, as well as probabilistic Turing machines.

REFERENCES

Arbib, M., 1965, Brains, Machines and Mathematics (McGraw-Hill, New York).

Ashby, R. W., 1958, An Introduction to Cybernectics (Chapman and Hall, London).

Bowen, R., 1978, On Axiom A Diffeomorphisms, Regional Conference Series in Math. No. 35 (Am. Math Society, Providence).

Bremermann, H., 1974 in: Conrad, Guttinger and Dal Cin (1974), 304.

Burks, J. W., 1970, Essays on Cellular Automata (University of Illinois, Urbana).

Conrad, M., 1972, Curr. Mod. Biol. (now BioSystems) 5, 1.

Conrad, M., 1974, J. Theor. Biol. 46, 167.

Conrad, M., 1976a, BioSystems 8, 119.

Conrad, M., 1976b, J. Neurosci. Res., 2, 233.

Conrad, M., 1978, in: Theoretical Approaches to Complex Systems, R. Heim and G. Palm (eds.), Springer Lecture Notes in Biomathematics No. 21 (Springer, Heidelberg).

Conrad, M. W. Guttinger and M. Dal Cin (eds.), 1974, Physics and Mathematics of the Nervous System, Springer Lecture Notes in Biomathematics, No. 4, (Springer, Heidelberg).

Efstratiadis, A., J. Rosakony, T. Maniatis, R. M. Lawn, C. O'Connell,
     R. A. Spritz, J. K. Deriel, B. G. Forget, S. M. Weissman et.
     al., 1980, Cell 21, 653.

Elliott, H. C. 1969, Textbook of Neuroanatomy (J. B. Lippincott,
     Philadelphia).

Garey, M. R. and D. S. Johnson, 1979, Computers and Intractability
     (Freeman, San Francisco).

Gilbert, W., 1978, Nature 271, 501.

Gould, S. J. and N. Eldredge, 1977, Paleobiology 3, 113.

Hastings, H. M. and R. Pekelney, 1982, BioSystems 15, 155.

Ho, M. W., and P. T. Saunders, J. Theor. Biol. 41, 535.

Hofstader, D. R. 1979, Gödel, Escher, Bach: An Eternal Golden
     Braid (Vinatge Books, New York).

Lashley, K. S., 1963, Brain Mechanisms and Intelligence (Dover,
     New York).

Lewin. R., 1980a, Evolutionary theory under fire. Science
     210, 883.

Lewin, R., 1980b, Is your brain really necessary. Science
     210, 1232.

Lin. C. C. and L. A. Segel, 1974, Mathematics Applied to Deter-
     ministic Problems in the Natural Sciences (McMillan, New York).

McCoy, B. M. and T. T. Wu, 1973, The Two Dimensional Ising Model
     (Harvand University Press, Cambridge, Mass.).

McCulloch, W. S. and W. Pitts., 1943, Bull. Math. Biophys. 5
     115.

Noda, H. 1982, J. Theor. Biol. 95, 145.

Ohno, S., 1970, Evolution by Bene Duplication (Springer, Berlin).

Ohno, S., 1972, in Smith (1972).

Paz, A., 1971, Introduction to Probabilistic Automata (Academic,
     New York and London).

Rosen, R. 1969, in Information Processing in the Nervous System,
     K. N. Leibovi (ed.) (Springer, Heidelberg).

Rössler, O. E., 1974, in: Conrad, Guttinger and Dal Cin (1974), 366.

Rössler, O. E., 1983, this volume.

Schmidt, R. F. (ed.), 1978, Fundamentals of Neurophysiology (Springer, New York).

Shannon, C. E., and W. Weaver, 1948, The mathematical theory of communication (University of Illinois, Urbana).

Shaw, R., 1983, this volume.

Smith, H. H. (eds.), 1972, Evolution of the Genetic Systems (Gordon and Breach, New York).

Smith, J. Maynard, 1970, Nature, London 225, 563.

Smith, J. Maynard and G. R. Price, 1973, Nature, London 246, 15.

Taylor, J. G., 1974, in: Conrad, Guttinger, and Dal Cin (1974), 230.

Totafurno, J., C. J. Lumsden and L. E. H. Trainor, 1980, J. Theor. Biol. 85, 171.

Tsetlin, M. C., 1973, Automaton Theory and Modeling of Biological System (Academic, New York).

Turing, A. M., 1936, Proc. Lond. Math. Soc. Ser. 2, 42, 230.

DESIGN FOR A ONE-DIMENSIONAL BRAIN

Otto E. Rössler

University of Tübingen

West Germany

I. INTRODUCTION

Is it possible to put consciousness into a home computer?
The only way one may have any chance to accomplish this is by re-
ducing the dimensionality of the class of problems that the brain
to be designed must solve. This can indeed be done within the
context of a certain abstract theory.

In the following, this theory will be outlined and a way of
how to design the simplest system of the most complicated subclass
indicated.

II. DEDUCTIVE BRAIN THEORY

Deductive biology is that subdiscipline of theoretical biology
that is predictive. As a first-principles theory, it works even
in the absence of any biological knowledge. There exists a subset of problems in adaptation theory where this desirable situation
indeed holds true. Consider the given survival functional that is
a function of position in space (or, more generally, space-time).
Then as long as this functional is indeed given (and is not changed
by some kind of metabolic adaptation-which is the only way out),

145

any system subjected to this functional has only one way of increasing its survival: positional adaptation (choosing a more
optimal position along the survival functional). Only locomotion
that is not against the shape of the survival functional (may it
now be far from optimal or fully optimal) can possibly increase
survival. Thus, prediction is possible.

The simplest case is that of "smooth gradients" in space.
Bacteria live in such an environment. As it turns out, the
remarkable "decision making" observed in these creatures (Adler
and Tso, 1974) can be understood in these terms. Some quantitative predictions (double-sum structure of a set of subfunctionals
to be computed internally; Rössler, 1976) have yet to be verified
experimentally.

The next simplest case, after the bacterial locomotion control
problem,is that of organisms living in a nongradient environment.
These environments can be idealized as consisting of spatially
localized point sources of n different types. Moreover, visiting
these sources is subject to a certain temporal constraint: the
locomotive system must always make it toward another source
of the same type within a certain maximal time interval. Thus,
there is a maximal traveling radius for each newly filled tank,
so to speak, and there are n tanks and n types of filling stations.
This is a variant to the well-known traveling salesman problem.
Instead of having to minimize mileage, the salesman has to stay
within the bounds of n resettable hour glasses. This problem is
known under the name of the bottleneck traveling salesman (Garey
and Johnson, 1980, p. 212). The point is that it is just as hard
to solve optimally as the other. That is, it requires an exponential increase in computing complexity as the requisite survival
time, in a sufficiently large array of randomly scattered stationary
sources of n types, is increased linearly.

Thus, while this certainly is an artificial situation (with
only stationary sources, and only yes-or-no time limits assumed),

the "brain" required to solve it already exceeds the material powers of our universe (as all traveling-salesman types of problems do; Garey and Johnson, 1980). So if biological brains are the most sophisticated machines of the universe (which may or may not be true), a possible abstract reason has been found.

Incidentally, the present class of problems has something in common with chaos theory. In chaotic dynamical systems that are computer-implemented, there is also an exponentially growing price to pay for staying within an ε-bound of the single exact solution (cf. Shaw, 1981). It may be possible to imbed the single optimal trajectory of complexity theory among its neighbors in a metric which allows one to apply the theory of chaos-generating n-dimensional endomorphisms. But this is an aside.

So far, only the case of the single optimal solution was considered. Instead of the single free parameter assumed so far (maximum overall traveling time), it is more natural to use two: (1) allowed error rate per tank-filling interval (say, $\varepsilon=0.001$); (2) ratio of "maximal traveling radius with one tank filling" over "mean distance between filling stations of the same type" ($\alpha \leq 1$). The first of these two is related to the former single parameter. The presence of the second parameter now means that the problem becomes an optimal solution problem only as the second parameter approaches a certain maximum value of its own close to unity.

In this more realistic case, one has not a single admissible solution, but a whole range of equally adequate ("satisfying") ones. This whole set can now be differentiated once more in terms of the complexity of its members.

While this is too demanding a task to solve in full generality, it can be dealt with semiempirically. It is easy to set up a series of increasingly complex solutions (both mathematically and in terms of the minimal hardware requirements made concerning sensor, C. P. U., and actuators) that all belong to just a few categories. These categories are: random locomotion; successive single-goal directed

locomotion (discrete direction optimization); smooth direction optimization over all directions (scalar optimization, zero-depth
optimization); path optimization.

The finally remaining solution (as α approaches α_{max}) invariably
requires path optimization. Smooth direction optimization is quite
efficient already, however (Rössler, 1974). Those direction optimizing techniques are especially interesting that can be "upgraded"
toward acquiring the performance of a path optimizer without having
to be discarded in toto. (This is the biological principle of continuity of design wherever possible.)

As it turns out, every direction optimizer needs panoramic (simultaneously presenting) distance sensors by definition. These are
highly costly. Whenever a less costly solution (with "focusing type"
distance sensors) is adopted, in addition a "flight simulator" is
necessary for obtaining a pseudosimultaneous environmental representation (as at least required by the direction optimizer). The
combination of these two devices, however, automatically lends itself
to a new mode of functioning, path optimization, and this is free of
charge (Rössler, 1981).

Thus, one may speculate that while insect brains might be sequential locomotion control machines (discrete direction optimizers),
with a discontinuity of design setting them apart from the necessarily
more complicated smooth direction optimizers possibly used by lower
vertebrates (and higher molluscs, respectively), there might be a continuity of designs from there toward higher vertebrate brains in which
the performance of the direction optimizer (hypothalamus plus limbic
system) is upgraded toward that of a path optimizer of the pseudosimultaneous, simulational type (using paleo- and neocortex). For
a similar biological view, see Harth (]983).

At any rate, however, there exists an equivalence class of autonomous optimizers (steering a course through actual space) that includes biological brains. Whether this "minimal" function of biological brains is the only one that counts - or whether additional,

nonadaptive properties play a major role - is a question left open
by the present approach.

III.   ON THE WAY TOWARD A ONE-DIMENSIONAL BRAIN

A closer look at the combination of a direction optimizer with
a universal flight simulator (Rössler, 1981) reveals a number of
design features. Most important, there is an off switch that shuts
off the sensors and motors yet permits them to go on optimizing - in
pure simulation. Secondly, it must be possible to store the obtained
results on a short-term basis. Thirdly, the final result may be stored
away more permanently (into a passive storage device) by way of a parallel content addressing scheme that simply uses the n pertinent scalar
values of the subfunctionals of the direction optimizer as the address.

But this is still too crude a picture. Apart from the simulation
mode, there must also be an exploration mode. It differs from simulation only insofar as the system actually moves around - but not to
collect, only to probe. This is simulation without simulation, so to
speak. The subsequent genuine simulation then relies on the so obtained data and modifies them further.

This explanatory/simulational mode does not possess any external
source type to which it responds. It rather responds to uncertainty
(and certainty) in the way of a potential that at first is barely
localized but can be sharpened.

While all the other (exploitative) subfunctionals and their
corresponding potentials (Rössler, 1974) are sharply focused on their
corresponding external sources, in the absence of any limit to sensor
resolution and range, the present ones are at first nonlocalized. But
potentials of this type are already necessary if the idealizing assumption of an obstacle-free unlimited observation range is dropped.
In this case there must be subfunctionals that at first indiscriminate-
ly endow all directions with an attracting (or repelling) specific
weight which then suddenly collapses as the corresponding source comes
in sight. Thus, again only a slightly less idealized environment had

to be assumed in order to obtain the required elements free of charge.

It then turns out that such an optimizer is able to cope not only with stationary sources (as were assumed so far), but also with moving or inconstant ones. As a consequence, it is now in principle possible to remove all of the (2 or 3) space dimensions that were assumed so far while sticking to the overall optimization scheme. This simplification has the asset that everything comes into the range of a modest amount of hardware. On the other hand, the problem of obtaining an accurate conceptualization before starting to tinker is still as great as it was in the original 4-dimensional case.

IV. SOME SPECIFIC SUGGESTIONS

The block diagram for the 4-dimensional case (Rossler, 1981) is too complicated to allow for easy inclusion of the above modifications. A purely temporal intermediary version will have to contain the following specific features: a) There is a finite time depth for the pseudosimultaneous representation of the one-dimensional (purely temporal) environment. Along this simultaneity range, the $\underline{n}$ subfunctionals (Rössler, 1974) have a time-integrated representation. This makes trade-offs between subfunctionals across time possible. b) Once a former uncertainty is removed at a certain moment in time, a "pull-back operation" retroactively wipes out the corresponding uncertainty functional, also for former moments. c) Upon return from the simulation mode, there is an automatic correction among the (meanwhile nonattended to) other contents of the simultaneity pad for the time that was consumed during the simulation. d) Once convergence (toward stable groupings of environmental features and associated subfunctionals) has been achieved under the simulation - so that all uncertainty has disappeared - the now more or less invariant values of the subfunctionals can be used as a label for the automatic storing into (an retrieval from) passive memory of the corresponding subscene. e) The whole system works at several levels of temporal resolution (tolerance in the sense of Zeeman, 1965)

simultaneously. If the levels differ in resolution by a factor of
two, an arbitrary number of them requires only twice as much hardware as a single one. The number of finer resolutions available is
to decrease toward the end of the existence interval of the time window. Shift of information between levels is made possible in both
directions through the presence of appropriate interpolation (combined with smoothing and sharpening) modules or compression modules,
respectively. These modules, as well as the main modules, are identical over all resolution levels. Successive shifts of information
then lead to the generation of the "idealized" geometrico-topological
shapes of finitely many types. These shapes form the natural units
of classification available to the system.

In the end, what will come out if all of these constraints are
incorporated, under a careful double-checking with the nontruncated
original version in an at least local (piecemeal) fashion, is a system that among other things will show the property of being condi-
tionable in both an operant and nonoperant manner. And this even
though none of the notions of behavioristic psychology were employed
in its design.

V. CONCLUDING REMARKS

A question that surfaced at this meeting of physicists and
biologists was the issue of consciousness (or the presence of subjective experience in dynamical systems). The above-proposed machine
is a candidate for a (potentially self-reflecting) dynamical system
of nontrivial type. Just one remark is on line concerning this most
difficult (Cartesian) subject. If anticipatory dynamical systems
(Rosen, 1981) of the class outlined above should possess a subjective correlate when physically implemented, then the same dynamical
systems (possessing the same equations) should, when implemented in
a slightly different physical manner, still possess this correlate.
The "slightly different" physical way refers to the possibility of
replacing one dimension (time) by one more space dimension - since
no more than two space dimensions are in principle needed for the

writing of the above system (cf. Dewdney, 1979). This abstract possibility of having a subjective experience tied to a crystal is perhaps the most disturbing general implication of dynamical approaches to the brain.

As a last remark, it should be stressed that the present approach toward brain function is compatible with several recent attempts to describe or generate an artificial working memory (Kosslyn and Shwartz, 1977; Albus, 1979; Hoffman, 1980; Pellionisz and Llinás, 1982). The basic difference, apart from the lower dimensionality assumed, consists in the claim made that the major features of the present system can in principle be derived from first principles.

VI. SUMMARY

A deductive approach is presented which puts biological brains into a larger equivalence class of "autonomous optimizers." Some hints as to how to design a maximally simplified (1-D) member of the most sophisticated (path optimizing) subclass are given.

REFERENCES

Adler, J.A. and W.W. Tso (1974). Decision-making in bacteria: Chemotactic response of E. coli to conflicting stimuli. Science 184, 1292-1294.

Albus, J. (1979). A model of the brain for robot control. Byte 4 (9), 130-148.

Dewdney, A.K. (1979). J. Recreational Math. 12, 16.

Garey, M. and D. Johnson (1980). Computers and Intractability. New York-San Francisco: Freeman.

Harth, E. (1983). A neural sketch-pad. These Proceedings.

Hoffman, W.C. (1980). Subjective geometry and geometric psychology. Math. Modelling 1, 349-367.

Kosslyn, S.M. and S.P. Shwartz (1977). A simulation of visual imagery. Cognit. Sci. 1, 265-295.

Pellionisz, A. and R. Llinás (1982). Tensor theory of brain
    function: The cerebellum as a space-time metric. Springer
    Lect. Notes in Biomath. 45, 394-417.

Rosen, R. (1981). On Anticipatory Systems. Monograph in press.

Rössler, O.E. (1974). Adequate locomotion strategies for an abstract
    organism in an abstract environment: A relational approach to
    brain function. Springer Lect. Notes in Biomath. 4, 342-369.

Rössler, O.E. (1976). Prescriptive relational biology and bacterial
    chemotaxis. J. Theor. Biol. 62, 141-157.

Rössler, O.E. (1981). An artificial cognitive map system.
    Biosystems 13, 203-209.

Shaw, R. (1981). Strange attractors, chaotic behavior and informa-
    tion flow. Z. Naturforsch. 36 A, 80.

Zeeman, E.C. (1965). Topology of the brain. In: Mathematics
    and Computer Science in Biology and Medicine (Medical
    Research Council Publication), pp. 277-292. London:
    Her Majesty's Stationary Office.

PROGRAM

20th ANNUAL ORBIS SCIENTIAE

DEDICATED TO P. A. M. DIRAC'S 80th YEAR

MONDAY, January 17, 1983

Opening Address and Welcome

SESSION I-A:	GENERAL RELATIVITY AND GAUGE SYMMETRY
Moderator:	Behram N. Kursunoglu, University of Miami
Dissertators:	P. A. M. Dirac, Florida State University
	Behram N. Kursunoglu, University of Miami

SESSION I-B:	RELATIVISTIC COSMOLOGY
Moderator:	Michael Turner, University of Chicago
Dissertators:	Edward W. Kolb, Los Alamos National Laboratory "COSMOLOGICAL AND ASTROPHYSICAL IMPLICATIONS OF MAGNETIC MONOPOLES"
	Michael Turner, University of Chicago "THE INFLATIONARY UNIVERSE: EXPLAINING SOME PECULIAR NUMBERS IN COSMOLOGY"
Annotators:	Pierre Ramond, University of Florida
	Anthony Zee, University of Washington

SESSION II:	ROUND TABLE DISCUSSION ON DIRAC'S CONTRIBUTIONS TO PHYSICS
Moderator:	Abraham Pais, The Rockefeller University
Dissertators:	Harish-Chandra, Institute for Advanced Study, Princeton
	Fritz Rohrlich, Syracuse University "THE ART OF DOING PHYSICS IN DIRAC'S WAY"
	Victor F. Weisskopf, Massachusetts Institute of Technology

155

TUESDAY, January 18, 1983

SESSION III-A:	CURRENT STATUS OF EXPERIMENTS (W's, Z^0, MONOPOLE, PROTON DECAY)
Moderator:	Maurice Goldhaber, Brookhaven National Laboratory
Dissertators:	Frederick Reines, University of California, Irvine, and Daniel Sinclair, University of Michigan "A SEARCH FOR PROTON DECAY INTO $e^+\pi^0$ IRVINE-MICHIGAN-BROOKHAVEN COLLABORATION"
Moderator:	Joseph E. Lannutti, Florida State University
Dissertators:	Alfred S. Goldhaber, State University of New York at Stonybrook "MAGNETIC MONOPOLES AND NUCLEON DECAY"
Annotator:	P. A. M. Dirac, Florida State University
SESSION III-B:	ROUND TABLE DISCUSSION OF FUTURE ACCELERATORS
Moderator:	Alan D. Krisch, University of Michigan
Dissertators:	V. Soergel, DESY, West Germany "HIGH-ENERGY ep FACILITIES" R. R. Wilson, Columbia University "VERY-HIGH-ENERGY pp FACILITIES" Nicholas P. Samios, Brookhaven National Laboratory "HIGH-LUMINOSITY pp FACILITIES"
Annotator:	Ernest D. Courant, Brookhaven National Laboratory
SESSION IV:	COMPARISON OF THEORY AND EXPERIMENT
Moderators:	Sydney Meshkov, University of California, Los Angeles Pierre Ramond, University of Florida
Dissertators:	Sydney Meshkov, University of California, Los Angeles "GLUEBALLS" L. Clavelli, Argonne National Laboratory "ON THE MEASUREMENT OF α_s" Pierre Ramond, University of Florida "B-L VIOLATING SUPERSYMMETRIC COUPLINGS" Gabor Domokos, Johns Hopkins University "SPONTANEOUS SUPERSYMMETRY BREAKING AND METASTABLE VACUA"

SESSION VII: THE HISTORY AND FUTURE OF GAUGE THEORIES

Moderator: M. A. B. Beg, The Rockefeller University

Dissertators: M. A. B. Beg, The Rockefeller University
                      "DYNAMICAL SYMMETRY BREAKING: A STATUS REPORT"

                      L. A. Dolan, The Rockefeller University
                      "KALUZA-KLEIN THEORIES AS A TOOL TO FIND NEW
                      GAUGE SYMMETRIES"

                      G. Lazarides, The Rockefeller University
                      "FLUX OF GRAND UNIFIED MONOPOLES"

                      S. Mandelstam, University of California,
                       Berkeley
                      "ULTRA-VIOLET FINITENESS OF THE N = 4 MODEL"

FRIDAY, January 21, 1983

SESSION VIII: GENERAL RELATIVITY IN ASTROPHYSICS

Moderators: P. G. Bergmann, New York University
                      Joseph Weber, University of Maryland and
                       University of California, Irvine

Dissertators: Giorgio A. Papini, University of Regina, Canada
                      "GRAVITATION AND ELECTROMAGNETISM COVARIANT
                      THEORIES A LA DIRAC"

                      Joseph Weber, University of Maryland and
                       University of California, Irvine
                      "GRAVITATIONAL WAVE EXPERIMENTS"

                      Kip S. Thorne, California Institute of
                       Technology
                      "BLACK HOLES"

Annotator: John Stachel, Boston University

SESSION IX: GENERAL RELATIVITY IN PARTICLE PHYSICS,
                      QUANTUM GRAVITY

Moderator: Anthony Zee, University of Washington

Dissertators: Anthony Zee, University of Washington
                      "REMARKS ON THE COSMOLOGICAL CONSTANT PROBLEM"

                      John Schwarz, California Institute of Technology
                      "A NEW FORMULATION OF N=8 SUPERGRAVITY AND ITS
                      EXTENSION TO TYPE II SUPERSTRINGS"

                      Heinz Pagels, The Rockefeller University
                      "GRAVITATIONAL GAUGE FIELDS"

Pran Nath, Northeastern University
"SUPERGRAVITY GRAND UNIFICATION"

Annotator: Richard Dalitz, University of Oxford

WEDNESDAY, January 19, 1983

SESSION V: FUNDAMENTAL PHYSICAL MECHANISMS OF BIOLOGICAL
                      INFORMATION PROCESSING

Moderators: Gordon Shaw, University of California, Irvine
                      F. Eugene Yates, University of California,
                        Los Angeles

Dissertators: Gordon Shaw, University of California, Irvine
                      "INFORMATION PROCESSING IN THE CORTEX: THE
                      ROLE OF SMALL ASSEMBLIES OF NEURONS"

                      Robert Rosen, Dalhousie University, Canada
                      "INFORMATION AND CAUSE"

                      Michael Kohn, University of Pennsylvania
                      "INFORMATION FLOW AND COMPLEXITY IN LARGE-SCALE
                      METABOLIC SYSTEMS"

                      George Wald, Harvard University
                      "LIFE AND MIND IN THE UNIVERSE"

                      Sidney W. Fox, University of Miami
                      "PHYSICAL PRINCIPLES AND PROTENOID
                      EXPERIMENTS IN THE EMERGENCE OF LIFE"

Annotator: Erich Harth, Syracuse University

THURSDAY, January 20, 1983

SESSION VI: FRAMEWORK OF ANALYSES OF BIOLOGICAL INFORMATION
                      PROCESSING

Moderator: Michael Conrad, Wayne State University

Dissertators: Michael Conrad, Wayne State University
                      "MICROSCOPIC-MACROSCOPIC INTERFACE IN BIOLOGICAL
                      INFORMATION PROCESSING"

                      Harold Hastings, Hofstra University
                      "STOCHASTIC INFORMATION PROCESSING IN BIOLOGICAL
                      SYSTEMS II - STATISTICS, DYNAMICS, AND PHASE
                      TRANSITIONS"

                      Otto E. Rössler, University of Tübingen,
                        West Germany
                      "DESIGN FOR A ONE-DIMENSIONAL BRAIN"

PARTICIPANTS

Ahmed Ali
DESY, Hamburg, Germany

M. A. B. Beg
The Rockefeller University

Carl M. Bender
Washington University

Peter G. Bergmann
Syracuse University

John B. Bronzan
Rutgers University

Arthur A. Broyles
University of Florida

Roberto Casalbuoni
Istituto di Fisica Nucleare
 Florence, Italy

Lay Nam Chang
Virginia Polytechnic Institute
 and State University

L. Clavelli
Argonne National Laboratory

Michael Conrad
Wayne State University

John M. Cornwall
University of California
 Los Angeles

Ernest D. Courant
Brookhaven National Laboratory

Thomas Curtright
University of Florida

Richard H. Dalitz
Oxford University
 Oxford, England

Ashok Das
University of Rochester

Stanley R. Deans
University of South Florida

P. A. M. Dirac
Florida State University

Louise Dolan
The Rockefeller University

Gabor Domokos
Johns Hopkins University

Susan Kovesi-Domokos
Johns Hopkins University

Bernice Durand
University of Wisconsin

Loyal Durand
Fermi National Accelerator
 Laboratory

Robert W. Flynn
University of South Florida

Sidney W. Fox
University of Miami

Andre I. Gauvenet
Electricite de France, Paris, France

Alfred S. Goldhaber
State University of New York

Gertrude S. Goldhaber
Brookhaven National Laboratory

Maurice Goldhaber
Brookhaven National Laboratory

O. W. Greenberg
University of Maryland

Franz Gross
College of William and Mary

Bernard S. Grossman
The Rockefeller University

Gerald Guralnik
Brown University

M. Y. Han
Duke University

Harish-Chandra
Institute for Advanced Study

Erich Harth
Syracuse University

Harold Hastings
Hofstra University

K. Ishikawa
City College of the CUNY

Gabriel Karl
University of Guelph

Boris Kayser
National Science Foundation

Michael S. Kohn
University of Pennsylvania

Edward W. Kolb
Los Alamos National Laboratory

Alan D. Krisch
University of Michigan

Behram N. Kursunoglu
University of Miami

Joseph E. Lannutti
Florida State University

G. Lazarides
The Rockefeller University

Y. Y. Lee
Brookhaven National Laboratory

Don B. Lichtenberg
Indiana University

Stanley Mandelstam
University of California
 Berkeley

Philip Mannheim
University of Connecticut

Jay N. Marx
Lawrence Berkeley Laboratory

Koichiro Matsuno
University of Miami

Sydney Meshkov
University of California
 Los Angeles

A. J. Meyer II
The Chase Manhattan Bank, N. A.

Stephan L. Mintz
Florida International University

John W. Moffat
University of Toronto

Paul Mueller
University of Pennsylvania

Darragh Nagle
Los Alamos National Laboratory

Pran Nath
Northeastern University

John R. O'Fallon
University of Notre Dame

Heinz R. Pagels
The Rockefeller University

Abraham Pais
The Rockefeller University

William F. Palmer
Ohio State University

Giorgio A. Papini
University of Regina
 Regina, Canada

Arnold Perlmutter
University of Miami

Pierre Ramond
University of Florida

L. Ratner
Brookhaven National Laboratory

Paul J. Reardon
Brookhaven National Laboratory

Frederick Reines
University of California
  Irvine

Fritz Rohrlich
Syracuse University

Robert Rosen
Dalhousie University
  Halifax, Canada

S. Peter Rosen
National Science Foundation

Otto E. Rössler
University of Tübingen
  Tübingen, Germany

Nicholas P. Samios
Brookhaven National Laboratory

Mark Samuel
Oklahoma State University

John H. Schwarz
California Institute of
  Technology

Joel Shapiro
Rutgers University

Gordon Shaw
University of California
  Irvine

P. Sikivie
University of Florida

Daniel A. Sinclair
University of Michigan

George A. Snow
University of Maryland

V. Soergel
DESY, Hamburg, Germany

John Stachel
Boston University

Jerry Stephenson
Los Alamos National Laboratory

Katsumi Tanaka
Ohio State University

Michael J. Tannenbaum
Brookhaven National Laboratory

L. H. Thomas
National Academy of Sciences

Kip S. Thorne
California Institute of Technology

Michael S. Turner
University of Chicago

Giovanni Venturi
Universita di Bologna
  Bologna, Italy

George Wald
Harvard University

Kameshwar C. Wali
Syracuse University

Joseph Weber
University of Maryland

Victor F. Weisskopf
Massachusetts Institute of
  Technology

Robert R. Wilson
Columbia University

F. Eugene Yates
University of California
  Los Angeles

G. B. Yodh
University of Maryland

Frederick Zachariasen
California Institute of
  Technology

Cosmas K. Zachos
Fermi National Accelerator
  Laboratory

Anthony Zee
University of Washington

INDEX